Simulations in Bulk Solids Handling

Simulations in Bulk Solids Handling

Applications of DEM and other Methods

Edited by

Don McGlinchey

WILEY-VCH

The Editor

Prof. Don McGlinchey
Glasgow Caledonian University
Department of Mechanical Engineering
Cowcaddens Road
G4 0BA Glasgow
United Kingdom

Cover Image: © your/Shutterstock

Library of Congress Card No.: applied for

British Library Cataloguing-in-Publication Data
A catalogue record for this book is available from the British Library.

Bibliographic information published by the Deutsche Nationalbibliothek
The Deutsche Nationalbibliothek lists this publication in the Deutsche Nationalbibliografie; detailed bibliographic data are available on the Internet at <http://dnb.d-nb.de>.

Print ISBN: 978-3-527-35010-0
ePDF ISBN: 978-3-527-83591-1
ePub ISBN: 978-3-527-83592-8
oBook ISBN: 978-3-527-83593-5

Typesetting Straive, Chennai, India
Printing and Binding CPI Group (UK) Ltd, Croydon, CR0 4YY

C117762_190123

Contents

Preface

The field associated with Bulk Solids Handling has always been open to a variety of science and engineering disciplines, and important contributions have come from Chemical, Civil, and Mechanical Engineering and Physics – all have contributed to the growth in simulation methods and technologies. Arguably the most important development in the past 20 years has been the progress in discrete element modelling (DEM) which has reached wide industrial use through commercially available codes, along with two-fluid models and the integration with computational fluid dynamics and multibody motion codes. Options for the high-fidelity simulation of bulk solids are now open to engineers with access to small HPC facilities or a reasonable workstation, and relevant software, both open source and commercial, can be accessed by all via an internet link through web services and cloud computing.

This book is intended to be of interest both to practicing engineers who may have had little or no exposure to the simulation of bulk solids during their formal education, and to graduate and post-graduate researchers in this area to provide background and perhaps a springboard for new ideas and areas for investigation. There is wide (and growing) use of simulation in science and engineering research, and industry practice; therefore, it is essential for students and practitioners alike to develop an understanding of relevant simulation techniques and methods. Knowledge of the capabilities and limitations of these techniques will lead to a better understanding of their potential to solve problems on various bulk solids handling applications and suggest potential research direction. Although covering the fundamentals in some sections, the book is not introductory in the sense of a first text in bulk solids handling or the discrete element method. However, the mathematics used in this book is generally undergraduate-level calculus applied to the underlying physics involved in particle technology and bulk solids handling operations and should be accessible to graduate engineers.

A simulation of a bulk solids handling process or operation may be performed for different reasons, for example:

1. to reproduce in a virtual environment the behaviours of a physical system so that such behaviours may be 'recorded'.
2. to generate simulated behaviours which can be used as ideal behaviours of a perfectly well-behaved physical system to aid identify degraded performance or problems.
3. to predict problems, e.g. erosion, before they occur on a physical system to allow mitigating action.

4. to validate design or operational choices in the virtual environment before instantiating in a physical system.
5. as the basis for the development of a reduced order model to form a component of a Digital Twin of the process or operation.

The bulk solids handling operations and areas covered include chute design, belt conveying, pneumatic conveying, scoops and grabs, port side operations, pharmaceuticals, particle shape, and simulation calibration.

Chapter 1 summarises different DEM calibration philosophies, the most important DEM parameters of the most commonly used contact models, as well as different calibration tests to determine the DEM parameters efficiently and unambiguously.

Chapter 2 discusses the main relevant aspects of transfer chute design to minimise problems in practice and shows how simulations can be used to improve the design. Modelling and evaluation issues for typical transfer chute problems like the simulation of build-up, the determination of impact forces, and the prediction of belt mis-tracking, abrasive wear, and dust emission are presented.

In Chapter 3 Finite and Discrete Element Methods are described to calculate the static and dynamic belt tensions within the conveyor belt during starting, steady-state running, and stopping, with the aim to improve performance and reliability of the system.

Chapter 4 presents the developments in the use of coupled Discrete Element Method and MultiBody Dynamics (DEM–MBD) in particle-based systems related fields with a particular focus on the applications in design of bulk handling and processing equipment.

Chapter 5 outlines the fundamentals of gas–solids flow in pipelines and details three simulation methods, Single- and Two-Fluid Models and combined CFD–DEM. The principal theories and equations commonly available in commercial off-the-shelf and open source software packages are given.

Chapter 6 introduces different methodologies for computationally capturing erosive wear and particle breakage in pneumatic conveying. A detailed discussion of one-dimensional two-phase flow modelling of the industrial conveying pipeline for particle breakage and erosion prediction, including the algorithm for breakage and erosion modelling, is included.

Chapter 7 presents a brief introduction of DEM modelling, including principles, parameter calibration, and model validation, and its application in pharmaceutical processing. For this purpose, an overview of DEM modelling of several key unit operations in pharmaceutical manufacturing, including powder mixing, continuous blending, twin screw granulation, die filling and dry powder inhalation, is presented.

Chapter 8 covers recent code optimisation for the contact force calculation with arbitrary body shape, parallelisation performance, and discussion of results showing both deformable and rigid-body versions of the code in the application of cube packing.

I am deeply grateful to all the authors for their excellent contributions to this book which provides a width and depth of knowledge in one volume I haven't seen elsewhere.

Glasgow, Scotland, UK
21st November 2022

Professor Don McGlinchey

1

Calibration of DEM Parameters

Corné Coetzee[1] *and André Katterfeld*[2]

[1] *Stellenbosch University, Department of Mechanical and Mechatronic Engineering, Granular Materials Research Group, Joubert Street, Stellenbosch, 7600, South Africa*
[2] *Otto von Guericke University Magdeburg, Chair of Material Handling, Institute of Logistics and Material Handling Systems, Universitätsplatz 2, 39106 Magdeburg, Germany*

Abbreviations and Acronyms

AoR	Angle of Repose
BCA	Bulk Calibration Approach
CFD	Computational Fluid Dynamics
CoR	Coefficient of Restitution
CPU	Central Processing Unit
DEM	Discrete Element Method/Discrete Element Modelling
DMA	Direct Measurement Approach
FEM	Finite Element Method
GPR	Gaussian Process Regression
GPU	Graphical Processing Unit
GSMC	Generalised Surrogate Modelling-based Calibration
MPM	Material Point Method
PSD	Particle Size Distribution
UCT	Uniaxial Compression Test

1.1 Introduction

Mesh-based simulation techniques in continuum mechanics, such as the Finite Element Method (FEM) or Computational Fluid Dynamics (CFD), require the body or volume of interest to be discretised by a mesh. The mesh closely represents the real object with small differences or idealisations, mostly directly proportional to the size of the mesh. These approaches show the convergence of the results with mesh refinement. However, Lagrangian approaches, such as FEM, suffer from severe mesh distortion when the body experiences large deformation. In these cases, the

Simulations in Bulk Solids Handling: Applications of DEM and other Methods, First Edition.
Edited by Don McGlinchey.

solution can become unstable, and the results inaccurate. The so-called meshless continuum-based methods such as the Material Point Method (MPM) are capable of modelling larger deformation [1]. However, these methods still assume a continuum body and might still rely on a non-deforming mesh. As a result, these methods cannot model the discrete nature of granular materials such as mixing and segregation or single particles separated from the bulk of the material in a screening process, for example.

The Discrete Element Method (DEM) was developed by Cundall and Strack [2] in the 1970s in order to solve problems associated with rock mechanics. The potential of DEM was quickly recognised for research purposes in a number of areas such as physics, nanotechnology, chemical engineering, and materials handling. DEM is completely meshfree (or meshless) and can easily model the large deformation typically associated with the handling and flow of bulk granular materials (particulate matter). DEM can also model the discrete nature of the individual particles, during screening for example.

To use DEM to analyse the behaviour of bulk materials, for example, in conveyor systems, during transportation and storage and flow through processing equipment, an accurate simulation model should be generated. A DEM model should define the geometric properties of the particles, such as the size and shape distributions, as well as the geometry of any structure or equipment. The interaction or contact properties (particle–particle and particle–wall) also need to be defined, which is a major component of the modelling process. As in all the numerical simulation methods, the experience (know-how) and sometimes the art applied by the user are critical to define and create a model capable of producing the most accurate simulation results in the shortest possible time frame.

In a DEM model, the discretisation of the bulk material is directly related to the size of the considered particles, which have a significant influence on the behaviour of the modelled material. DEM is also computationally intensive, and for this reason, most practical models are simplified in terms of particle size, shape, and contact properties. This idealisation is the reason why established bulk material properties (e.g. the angle of internal friction and the angle of repose [AoR]) cannot be directly used as input parameters. Hence, it is necessary to reverse engineer the parameters by comparing the modelled bulk behaviour to that observed in the experimental tests. This procedure is called the 'calibration of DEM parameters' and is the key to produce realistic simulation results.

1.2 Basic DEM Theory

A typical DEM model consists of particles and walls. The particles can make contact with one another and with walls, but wall–wall contact is usually undefined. The particles represent the granular material and can in theory take on any shape and size. However, in practice, spherical and multi-sphere particles are the most commonly used. Walls are used to define all the structures with which the particles can interact, such as the walls of equipment and machines. Contact models are used to calculate the contact forces and moments based on the contact kinematics.

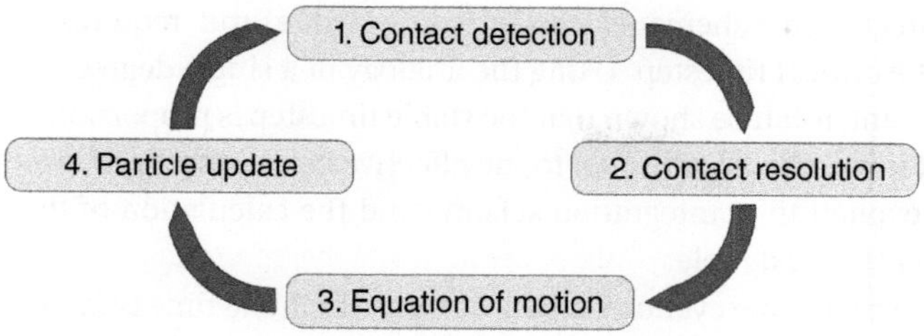

Figure 1.1 DEM computation cycle.

1.2.1 The Calculation Cycle

A DEM calculation cycle consists of four steps as illustrated in Figure 1.1, namely (1) contact detection, (2) contact resolution, (3) solving the equation of motion, and (4) updating of the particle velocity and position.

The contacting pieces (particles and walls) are allowed to overlap, and in the first step of the computation cycle, all particle–particle and particle–wall contacts are identified. The overlap is assumed to be relatively small compared to the particle size. Although contact detection happens automatically, without any user intervention, the particle shape selected by the user has a significant effect on the efficiency of this step. Spherical particles are computationally the best, followed by multi-sphere particles and more complex shape representations such as super-quadrics and polyhedra.

In the second step, the force and moment vectors are calculated at each contact, based on one of a number of available contact models selected by the user. The contact force and moment are dependent on the relative contact displacement or overlap (elastic force), velocity (viscous force) and the contact parameter values specified by the user.

In the third step, the resultant force and moment acting on each particle are calculated. This includes the forces, $\mathbf{F}_c$, and moments, $\mathbf{M}_c$, due to the contacts and the body force due to gravity $\mathbf{F}_g$. Based on the particle's mass, m, and moment of inertia, I_g, the translational acceleration, $\ddot{\mathbf{x}}$, and the rotational acceleration, $\ddot{\boldsymbol{\theta}}$, can be calculated using the equations of motion,

$$\ddot{\mathbf{x}} = \frac{\mathbf{F}_c + \mathbf{F}_g}{m} \tag{1.1}$$

$$\ddot{\boldsymbol{\theta}} = \frac{\mathbf{M}_c}{I_g} \tag{1.2}$$

In the fourth and last step, the particle velocity (translational $\mathbf{v} = \dot{\mathbf{x}}$ and rotational $\boldsymbol{\omega} = \dot{\boldsymbol{\theta}}$) is first updated using an explicit time integration scheme,

$$\mathbf{v}^{t+\Delta t} = \mathbf{v}^t + \ddot{\mathbf{x}}\Delta t \tag{1.3}$$

$$\boldsymbol{\omega}^{t+\Delta t} = \boldsymbol{\omega}^t + \ddot{\boldsymbol{\theta}}\Delta t \tag{1.4}$$

where Δt is the timestep. This is followed by the particle's position and orientation update,

$$\mathbf{x}^{t+\Delta t} = \mathbf{x}^t + \mathbf{v}^{t+\Delta t}\Delta t \tag{1.5}$$

$$\boldsymbol{\theta}^{t+\Delta t} = \boldsymbol{\theta}^t + \boldsymbol{\omega}^{t+\Delta t}\Delta t \tag{1.6}$$

The explicit time integration scheme is conditionally stable and requires a timestep smaller than the critical timestep. Using the analogy of a single degree-of-freedom mass-spring system, it can be shown that the stable timestep is proportional to the particle mass and inversely proportional to the effective contact stiffness. For slight variations in the explicit time integration scheme and the calculation of the timestep, see O'Sullivan [3] for example.

This step concludes the basic time cycling sequence, after which the time is incremented, followed by a new contact detection step.

1.2.2 Contact Models

A contact model defined at each contact describes the force–displacement relation. There are a number of contact models from which the user can select. Figure 1.2 shows the basic elements of a contact model, namely springs, dashpots, frictional sliders, and tension elements. A combination of these elements act in each of the normal and shear (tangential) directions.

The spring elements define the elastic force component and can have linear behaviour (as in the linear model) or non-linear behaviour (as in the Hertz–Mindlin model). The viscous dashpots dissipate energy, and the frictional slider allows for Coulomb-like frictional behaviour in the shear direction. For the modelling of spheres, rolling resistance models are very important; however, they are not visualised in Figure 1.2. Cohesive behaviour can be modelled by allowing tensile forces in the normal direction. The details of the different contact models are not presented here, and the interested reader should consult other sources such as O'Sullivan [3] and Thornton [4].

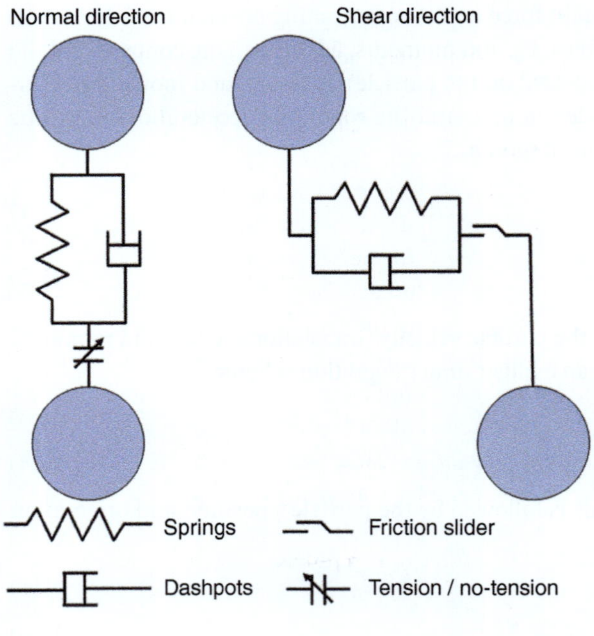

Figure 1.2 Typical elements of a contact model.

At each timestep, the relative motion between two contacting pieces is used, in combination with the elements defined above, to update the contact force components. The user should specify the parameter values for the spring stiffness, for example the damping (dashpot) constants and the coefficient of friction. Obtaining a set of parameter values so that the modelled bulk behaviour of the material is an accurate representation of a physical material is the main focus of the calibration process.

1.3 DEM Application and Calibration Philosophies

In general, the application of DEM can be classified in two groups:

- **Group 1** focuses on improving our general understanding of the physics (e.g. rheology, constitutive behaviour) of granular materials through applied research [5, 6]. For this purpose, idealised ('artificial' or manufactured) bulk materials such as glass beads with a very well-defined particle size distribution (PSD) (mono-disperse or bi-disperse) and (close to) homogeneous properties are used. Often the particle behaviour is investigated in small-scale laboratory tests with a relatively small number of particles involved. The DEM model replicates the real application with a scale of 1 : 1, even if it requires a very large computational effort and duration of time.
- **Group 2** focuses on the modelling of 'natural' bulk materials in industrial applications. The materials range from powders to rocks and require an idealisation of the DEM models in terms of particle shape, size distribution, and stiffness.

In the first group of applications, it is possible to use tests to determine the particle or contact micro-properties and directly use them as DEM input parameters. This approach or philosophy for DEM parameter selection was labelled the 'Direct Measurement Approach' (DMA) by Coetzee [7].

In the second group of applications, the DMA can at most be partly used, since there are a number of simplifications required. For these applications, the macroscopic (bulk) material behaviour needs to be replicated by an idealised DEM model. Hence, the microscopic DEM parameters are to be chosen in such a way that the macroscopic behaviour of the material in the simulation is the same as that in reality. According to Coetzee [7], this approach is called the 'Bulk Calibration Approach' (BCA) since the process of finding the DEM parameters is an iterative process that involves a series of simulations which replicate at least one experiment. This approach truly earns the name 'calibration', and in the sections to follow, the term calibration refers to BCA, unless stated otherwise.

Although the process of parameter selection using the BCA is essential to produce simulations with realistic results, there are only a few works which try to define a systematic approach for the BCA. Especially in the industrial sector, many calibration approaches do not earn the name 'calibration'. The selection of DEM parameters based on a non-systematic parameter selection ('guessing of parameters') may lead to wrong DEM results and invalid conclusions. This may also be the case when DEM parameters are 'tuned' based on simulations of the final application, where the parameters are non-systematically varied until the flow results look like what is expected.

Furthermore, the kind of applications being modelled can also influence the calibration experiments and DEM parameters to be used (the selection of appropriate contact models, for example). Here the focus is on bulk storage and handling applications where the consolidation pressure is relatively low (less than 10 kPa), and the material is either static (storage in a bin or moving with a conveyor belt, for example) or dynamic and flowing (out of a bin or through a transfer chute, for example). When the material is under higher consolidation pressure, aspects such as plasticity might become important, which is not addressed in this work.

To ensure that the DEM simulation results can be trusted, the user must have a basic understanding of the physical bulk properties and the most important DEM parameters. This is addressed in the next sections, followed by a detailed discussion of the calibration process.

1.4 Physical Bulk Properties

The properties that are important for bulk handling applications are discussed. It is important to consider all these bulk properties during the calibration process.

1.4.1 Bulk Density and Porosity

Bulk density is needed to accurately model the body forces due to gravity and hence the consolidation pressure (even if relatively low) experienced by a bed of material. Bulk density is also important to accurately model the forces exerted by the material on structures or equipment walls. When the material flow rates are considered (modelling the conveying of material, for example), it is important to accurately model not only the mass flow rate but also the volume flow rate. The cross-sectional area of a transfer chute, for example, should be sufficient to handle the volume flow rate, which is related to the mass flow rate by the bulk density.

1.4.2 Bulk Friction

In this chapter, the term 'bulk friction' is used to define and characterise the bulk property related to the material's resistance to shear flow. Physically, the bulk friction is influenced by the PSD, particle shape, and inter-particle sliding friction. The particle shape leads to mechanical interlocking, which results in shear resistance even under zero normal load. The Coulomb-type sliding friction at contacts has a significant effect on the bulk friction. On the bulk level, the material can be considered a continuum and described by the Mohr–Coulomb constitutive model, defining a bulk cohesion and internal friction angle (yield locus) which can be measured in a direct shear test, for example.

1.4.3 Dissipation of Energy

In non-cohesive bulk materials, energy is dissipated in the form of contact friction, inelastic deformation of the particles and walls, particle breakage (fracture energy), and other losses such as wind resistance. Losses due to wind resistance and particle breakage are usually ignored, unless multiphase flow (coupled CFD–DEM modelling) or particle breakage is specifically analysed. Contact friction is responsible for the majority of the energy dissipated and should be accurately accounted for in all DEM models.

Additional contact damping (usually viscous damping) is required to reach a state of static equilibrium, when particles are dropped from a height into a bin, for example. However, appropriate levels of damping are also required for dynamic processes to keep the energy levels realistic and to avoid excessive particle motion.

1.4.4 Bulk Stiffness

When a bed of material is compressed, the relation between the applied force and the displacement (change in bed height) defines bulk stiffness. The stiffness can be measured under confined and unconfined conditions (these two approaches will result in different stiffness values).

Even if the inter-particle contact stiffness is linear, bulk stiffness is not necessarily linear. The non-linearity in bulk stiffness is caused by re-organisation of the position of the particles and contacts in the packing – increasing the number of contacts under increasing load. Due to this phenomenon, initial cyclic loading-and-unloading compression of a loosely packed particle bed results in hysteretic behaviour, where the unloading curve shows increased stiffness as compared to the loading curve.

1.4.5 Bulk Cohesion and Adhesion

In general, the term 'cohesion' refers to the attractive force between two similar materials, and the term 'adhesion' to the attractive force between two dissimilar materials. Here, however, the two terms are used interchangeably. The bulk cohesive behaviour is due to cohesive forces acting at the contact level (particle–particle and particle–wall), where it is mainly caused by one of two mechanisms: liquid-bridges or Van der Waals forces.

When moisture or liquid (mostly water in the applications considered here) is introduced, a thin liquid film forms on the outer surface of the particles and walls. When a new contact forms, a liquid bridge is formed, resulting in a tensile force due to the capillary effect and liquid surface tension (see Mitarai and Nori [8], for example). The van der Waals force, on the other hand, acts between two macroscopic bodies due to intermolecular cohesion forces, and the effects are significant only for very small particles such as powders (<100 μm, [9]). Independent of the cohesion mechanism at the contact level, an increase in contact cohesion results in an increase in bulk cohesion. Here, the term bulk cohesion refers to the bulk material's resistance to shear flow under zero normal load.

1.5 DEM Parameters and Their Relation to Bulk Properties

The DEM contact parameters discussed in this section are generic and applicable to all non-cohesive contact models, including linear and non-linear models. These are the parameters that should in general be considered and calibrated to ensure accurate modelling of the applications considered in this chapter. Although the implementation of these generic parameters is dependent on the specific contact model (and even the software) used, the relation with bulk properties as described here will remain valid.

The DEM parameters and bulk properties considered are listed in Table 1.1. The first three parameters (particle shape, size, and density) are related to the particles, while the other parameters are all related to the contacts (particle–particle and particle–wall). The relation between each parameter and bulk property (as discussed in Section 1.2) is indicated as either 'strong', 'weak', or 'negligible' (insignificant), and the effect of the parameter on the computation time is also indicated. These relations are further discussed below, and typical calibration experiments for each bulk property are listed at the bottom of the table.

1.5.1 Particle Shape

Discrete elements are used in DEM to represent the particles. In general, these elements are referred to as particles, and pioneering DEM codes [2] were used for circular (in 2D) and spherical (in 3D) particles due to the associated computational efficiency in contact detection and overlap calculation. However, with advances in computing power, more complex non-spherical shapes were introduced using different techniques as discussed below.

The particle shape will influence all the other model parameters that need to be calibrated and should be the first parameter to be decided by the modeller. If, at any later stage, the shape is changed or even slightly modified, all the parameter values should be re-calibrated. When the physical particles are non-spherical (which is the case in most industry and practical applications), the user has the option to make use of spherical particles and include rolling resistance or to make use of one (or more) of the available non-spherical shape models (with or without rolling resistance).

1.5.1.1 Non-spherical Particles

Any number of spheres can be merged to form a single particle, often referred to as 'multi-sphere particles', 'clumps', 'clusters', or 'glued particles'. The constituent spheres can overlap, and their relative position within the particle remains fixed, thus creating a rigid particle. Multi-sphere particles still allow for efficient contact detection and overlap calculation. However, these particles still have some limitations, and sharp edges or blocky particles cannot be easily created. Multi-sphere particles also have 'bumpy' surfaces, which might introduce higher levels of interlocking and tend to have more contacts with other particles and walls.

Table 1.1 Relation between DEM parameters and non-cohesive bulk properties.

DEM parameters	Bulk material properties: Bulk density and porosity including changes due to flow (dilatancy)	Bulk friction/shear/ interlocking/flow behaviour	Dissipation of energy (damping)	Bulk stiffness	DEM model computation time
Particle shape	**Weak** Influences the packing porosity, and hence the bulk density for a given particle (solid) density.	**Strong** The more non-spherical the shape, the higher is the interlocking effect, and thus, the higher the bulk friction.	**Negligible**	**Negligible**	**Strong** Spherical particles are the most efficient. The more complex the shape, the longer a defined simulation takes to complete.
Particle size distribution (PSD)	**Weak** Influences the packing porosity, and hence the bulk density. The wider the PSD, the lower is the porosity and the higher is the bulk density.	**Weak** The wider the PSD, the lower is the porosity and the higher is the particle interlocking, thereby increasing the resistance to shear.	**Negligible**	**Negligible**	**Strong** A reduction in the number of particles, logarithmically reduces the computation time.
Particle density	**Strong** The relation between particle density and bulk density is very close to linear for a given porosity (i.e. particle shape and size distribution).	**Negligible**	**Negligible**	**Negligible**	**Strong** The time step increases in size with an increase in density, thus reducing the total simulation time.

(Continued)

Table 1.1 (Continued)

	Bulk material properties				
DEM parameters	**Bulk density and porosity including changes due to flow (dilatancy)**	**Bulk friction/shear/ interlocking/flow behaviour**	**Dissipation of energy (damping)**	**Bulk stiffness**	**DEM model Computation time**
Contact damping/ coefficient of restitution	**Negligible**	**Negligible**	**Strong** Contact damping (together with sliding and rolling friction) is one of the major mechanisms for the dissipation of energy.	**Negligible**	**Negligible**
Contact stiffness	**Negligible** Using realistic values, the contact stiffness has a negligible effect on the bulk density. Very low values will increase the contact overlap and the bulk density. However, if this is the case, the stiffness is probably too low for modelling the typical materials considered here.	**Negligible** The contact stiffness has an effect on the bulk friction only for unrealistically low stiffness values which should be avoided [10].	**Negligible** If the user specifies either the non-dimensional critical damping ratio or coefficient of restitution, the effect is negligible since the code will calculate the damping constant based on this value in combination with the stiffness.	**Strong** The contact stiffness has a strong relation to the bulk stiffness. With an increase in the contact stiffness, the bulk stiffness increases.	**Strong** The contact stiffness highly influences the time step: the higher the stiffness, the smaller is the time step. Due to this effect, most DEM simulations make use of a reduced stiffness.

Contact coefficient of sliding friction: particle–particle	**Weak** Higher friction values lead to less dense packings (slightly higher porosity).	**Strong** The sliding friction is one of the major parameters influencing the bulk shear behaviour. An increase in the friction increases the bulk friction asymptotically: for lower friction coefficients, the bulk friction is more sensitive than for higher friction coefficients (typically 0.7 and higher).	**Strong** The sliding friction is one of the major mechanisms for energy dissipation.	**Negligible**	**Negligible**
Contact coefficient of sliding friction: particle–wall	**Negligible**	**Strong** The sliding friction is one of the major parameters influencing the bulk shear behaviour against a structure or wall. For example the flow down a chute or out of a bin.	**Strong** The sliding friction is one of the major mechanisms for energy dissipation, although quantitatively less than particle–particle friction.	**Negligible**	**Negligible**

(Continued)

Table 1.1 (Continued)

	Bulk material properties				
DEM parameters	**Bulk density and porosity including changes due to flow (dilatancy)**	**Bulk friction/shear/ interlocking/flow behaviour**	**Dissipation of energy (damping)**	**Bulk stiffness**	**DEM model Computation time**
Contact coefficient of rolling friction (particles and walls)	**Negligible**	**Strong** The rolling friction, in combination with the sliding friction, has a strong influence on the bulk friction (shear) behaviour.	**Negligible**	**Negligible**	**Negligible**
Experiments that are sensitive to the specific bulk material property	Bulk density/porosity: Filled container Dilatancy: Direct shear test	Angle of repose Discharge test Draw down test Direct shear test	Drop test Pendulum test Uniaxial compression	Uniaxial compression	—

Non-spherical particle shapes can also be modelled using mathematical descriptions such as superquadrics, polyhedra, and faceted particles. The advantages of this technique include the modelling of complex shapes more accurately, including blocky particles and sharp edges; however, it comes with a decrease in computational efficiency.

In general, when non-spherical particles are used, rolling resistance (friction) can be omitted if the selected shape can be accurately calibrated using only sliding friction [11].

1.5.1.2 Spherical Particles

The contact model for spherical particles should include rolling resistance to realistically model the rotational behaviour of physical non-spherical particles. Various authors (e.g. [11–14]) have shown that rolling resistance can accurately account for the particle shape in terms of bulk behaviour. Wensrich et al. [15] showed that spherical particles with rolling friction can accurately model the behaviour of non-spherical particles in both static and dynamic flow conditions with the dilational nature also captured to a reasonable degree. Several rolling resistance models are available as summarised by Ai et al. [16], and according to Wensrich and Katterfeld [17], Type C is the preferred and most stable model.

1.5.1.3 Shape Selection

If the bulk material has a relatively homogeneous particle size and shape, it is suggested to make use of spherical particles with rolling resistance. This is applicable to materials such as sand, crushed gravel, mineral ore, and agricultural grains and seeds. Also, when the material flow is relatively homogeneous, i.e. the material flows in a stream with little variation in particle velocity, the effect of particle shape becomes less significant, and a simple and efficient shape can be selected. In these cases, if non-spherical particles are used, it is proposed to use simple multi-sphere particles comprising three spheres in a pyramid shape [18, 19], where rolling resistance can be ignored.

In applications where the material is non-homogeneous in particle shape and size and where the particles are relatively large in comparison to the model boundaries (interacting with structures, equipment, and machine parts), the particle shape becomes more important. Also, if mixing, screening and sieving, or mechanical arching and bridging are investigated, accurate shape models should be considered [20]. The modelling of materials such as biomaterials might also require more accurate shape modelling if the particles are far from spherical, for example, wood chips and fibres [21, 22].

1.5.2 Particle Size Distribution and Scale

The PSD and scale (size) of calibration experiments are closely related. The PSD of the physical material being modelled can be measured using sorting sieves or screens. Alternative methods include photo analysis and laser diffraction techniques. However, the modeller should decide how the PSD of the physical material will be modelled. Due to computational constraints, the maximum number of particles that can be modelled within a reasonable time frame is limited, depending on the available software and hardware (Central Processing Unit [CPU] versus Graphical Processing Unit [GPU], for example). In most industrial-scale applications the number of particles can easily be in billions to trillions [23], which cannot be modelled, regardless of the available hardware. In these cases, the number of particles should be reduced by scaling the PSD.

There are several approaches available for this, as described by Roessler and Katterfeld [24] and the references therein, namely exact scaling, coarse graining, scalping, and combined scaling. Mohajeri et al. [25] developed a hybrid scaling technique where the geometry and the particles are scaled in a stepwise manner. This technique requires more simulation runs, but it can be used for the calibration and validation of upscaled particles even if the available experimental setup is too small to accommodate the final upscaled particles and if the experiment is scale-dependent.

The final selection of the modelled PSD will depend on the physical PSD, the application being modelled, available calibration equipment, and the available computational power. There are no general rules and guidelines that apply to all possible cases. The PSD used in the calibration step should be identical to the PSD used in modelling the (full-scale) final application. Similar to the particle shape, once a PSD is calibrated, it cannot be modified, unless the calibration steps are repeated. As a general rule, the particle size should be selected so that there are at least 10–20 particles across all dimensions of the geometry. The particle resolution should be high enough to render realistic and accurate results within a reasonable time frame. With larger (fewer) particles, the resolution is reduced, and boundary (wall) effects become significant. When the flow of the material through an orifice is modelled, the ratio of the orifice opening to particle size should be as large as computationally possible to avoid any choking effects.

1.5.3 Particle Density

Despite the idealised PSD and particle shape, the bulk density can still be accurately modelled by calibrating the particle (solid) density. The best approach is to follow the technique often used to experimentally measure the material's bulk density. A container with known volume is filled with modelled particles, and the total particle weight is determined and used to calculate the modelled bulk density. Iteratively changing the particle density, the modelled bulk density will converge to the measured value (allow for static equilibrium to be reached after each change in particle density).

This process is suitable for the relatively low consolidation pressures (less than 10 kPa) and non-cohesive to slightly cohesive materials as considered in this chapter. For higher consolidation pressures and cohesive materials, the density test should be performed with different consolidation pressures or at least in the same range of pressures that the material will be subjected to in the final application.

1.5.4 Contact Damping

The dissipation of energy is accounted for by frictional contacts and by introducing viscous damping at the contacts. Viscous damping is modelled as a dashpot in parallel to the contact stiffness (spring). In general, depending on the specific software implementation, the level of damping can be specified in one of three ways:

1. Directly providing the damping coefficient in units of $\mathrm{N\,s\,m^{-1}}$
2. Specifying the dimensionless critical damping ratio, ζ
3. Specifying the dimensionless coefficient of restitution (CoR)

1.5.5 Contact Stiffness

In DEM, the stiffness in the normal and tangential directions is used to calculate the contact forces. The formulation of the stiffness depends on the specific contact model used, and commonly used models include the Hertz–Mindlin (no-slip) model and the linear model.

In the linear contact model, the normal stiffness and tangential stiffness are independently specified in units of $\mathrm{N\,m^{-1}}$, i.e. linear spring elements. In the Hertz model, the particle's elastic properties are specified, either the Young's modulus or shear modulus, and Poisson's ratio. Based on these properties, the non-linear normal stiffness and tangential stiffness are calculated, which is a function of the overlap.

With the Hertz–Mindlin contact model, realistic bulk flow results can often be achieved using relatively low values for Young's modulus, starting at 1e7 Pa [24, 26, 27], and calibration is not necessary and often not performed. For the linear contact model, a lower-end contact stiffness is in the order of 1e4 $\mathrm{N\,m^{-1}}$. If forces or pressures have to be predicted, it must be determined how the reduction of the Young's Modulus will influence the simulation results, see for example, the detailed description by [27].

1.5.6 Particle–Particle Sliding Friction

The particle–particle coefficient of friction considered here cannot be measured directly due to their shape and size. But even if it could be measured, the shape and size are almost always simplified for computational efficiency. The whole calibration process compensates for these simplifications, and in the end, the combination of particle shape, size, sliding, and rolling friction has to produce realistic bulk flow behaviour. Thus, the final value used for the coefficient of sliding friction is not necessarily equal to that between two physical particles of the same material.

1.5.7 Particle–Wall Sliding Friction

The particle–wall sliding coefficient of friction can be directly measured in wall shear tests or inclined wall (plane) tests. In the applications considered here, the modelled bulk behaviour is usually less sensitive to the particle–wall friction than to the particle–particle friction.

1.5.8 Contact Rolling Friction

The rolling friction of non-spherical particles cannot be directly measured. Often different values are used for the rolling friction of particle–particle and particle–wall contacts. However, due to the fact that the rolling friction compensates primarily for the effect of the particle shape, the same value can be used for both. When spherical particles are used, rolling friction should be included, but when non-spherical particles are used, the inclusion of rolling friction will depend on the exact shape used [11].

1.6 Overview of Calibration Tests

The main objective of the calibration process is to compare the results from an experiment and a simulation, from which the DEM parameters are then derived, which will provide the most realistic and accurate results. Most calibration experiments need to consider the macroscopic (bulk) material behaviour, i.e. the behaviour of the interaction of a large number of particles (BCA). However, there are a small number of DEM parameters for which it is sufficient to investigate single particle/contact behaviour (DMA, e.g. a drop test to determine the CoR).

Standard tests for measuring bulk material properties include direct shear tests, using either translational (Jenike) or rotational (Schulze) shear testers. However, due to the complexity and the time duration of these tests, other non-standard tests were developed.

Katterfeld et al. [28] proposed in a White Paper that the calibration experiment should be chosen according to the real bulk material's flow mechanism in the final (industrial) application. If, for example, the final aim is to optimise the design of a transfer chute with a relatively fast flow regime and low to medium consolidation (typically $\ll 100$ kPa), the calibration experiment should consider the same flow regime. If compaction or comminution processes are analysed, a different set of calibration tests might be more appropriate.

Due to the very large variety of bulk solid handling applications, where DEM simulations may be of interest, it is not possible to describe each calibration experiment. The following calibration tests, each having advantages and disadvantages, can be used to study applications with fast flow regimes and low to medium consolidation as mentioned above. Coetzee [11] showed that the results of a proper calibration procedure can be reproduced by a different calibration procedure by using different experiments.

1.6.1 Basic Requirements for Calibration Tests

Due to the many possibilities, it is beneficial to define some practical evaluation criteria for the calibration tests. A typical test should:

- be simple and quick to undertake, both experimentally and numerically (simulation).
- Require only a small amount of bulk material.
- Provide a test result that can be easily measured or determined (experimentally and in the simulation).
- Be scale invariant. This means that the procedure should provide the same test results independent of the physical size of the test apparatus. The determination of an angle (such as the AoR for example) is typically invariant to the size of the particles and the test setup, as long as the flow mechanism which leads to the formation of the angle, is not significantly influenced by the size of the setup used [24].
- Produce consistent results. Due to the discrete (stochastic) nature of granular materials, two identical experiments will not necessarily provide identical results. Experiments should be repeated, and the results statistically analysed to obtain at least an average value and the standard deviation. Ideally, the experiment should produce consistent results, with a relatively small standard deviation.

All these tests do not necessarily result in the same AoR. This is mainly due to differences in the particle kinetic energy during the execution of the test, which influences the settling of the particles. Therefore, it is also important to simulate the exact same test and not to directly compare the AoR from one test apparatus to that of another.

All tests typically allow for an optical measurement of the AoR. The angle should be measured on a part of the slope where it is constant (approaching a straight line), and on several cross sections, so as to obtain an average value. Several image processing and measurement algorithms have been developed to ensure a consistent technique is applied and to eliminate human error in measuring the angle. A number of algorithms for experimental and simulation results can be found in Wensrich and Katterfeld [17], Tan et al. [30], Muller et al. [31], and Klanfar et al. [32].

1.6.2 Calibration Tests

Table 1.2 provides a brief summary and graphic representation of the most commonly used calibration tests from multiple studies. Since it is one of the most widely considered bulk material calibration tests (and parameter), the only test which should be explained in more detail is the AoR. To determine the AoR, several tests can be undertaken [31, 47]:

- Lifting cylinder test
- Funnel test (also called the hopper or drained test)
- Shear box (also called the ledge, slump, or rock box test)
- Cone test (also called the hourglass, sandglass, or trapdoor test)
- Rotating drum test (also called the dynamic AoR test)

Table 1.2 A summary and graphic representation of typical calibration tests.

Calibration test		Test properties			Comments and further references
		Scale effects	**Bulk measures for calibration**	**Test effort, applicability**	
Angle of repose tests	Lifting cylinder test 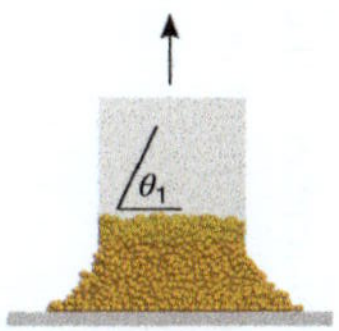	Scale invariant if the cylinder speed is adopted according to the scale.	The static AoR.	The test apparatus is very simple and easy to manufacture if not readily available (open-ended cylinder). The tests are simple and quick to execute (simulations and experimentally). Suitable for all PSD's if the cylinder diameter is selected to accommodate enough of the largest particles across its diameter.	The AoR is sensitive to the cylinder lifting speed. This effect can be significantly reduced if a relatively low speed is used. For further information see: Roessler and Katterfeld [29], Tan et al. [30], Muller et al. [31], Klanfar et al. [32], Qi et al. [33], Zhou et al. [34], Do et al. [35], Derakhshani et al. [36] and Elskamp et al. [37].
	Funnel test 	Scale variant due to the change in drop height.	The poured AoR.	The test apparatus is simple and easy to manufacture. The tests are simple and quick to execute (simulations and experimentally). Suitable for all sample and PSD sizes if the orifice is selected to accommodate enough of the largest particles across its diameter.	The AoR is sensitive to the drop height. The discharge time can be used as an additional calibration parameter (see the discharge test below). The formation of large-scale stockpiles can be seen as an industrial equivalent and a large-scale variant of this test. Due to the higher induced energy in the system, the AoR is usually lower than in the lifting cylinder test. Also, the difference in the AoR from different bulk materials is smaller than what it would be in the lifting cylinder test. Hence, if the measurement error is considered, the AoR measured in this test spans the same range for a variety of different materials. For further information see: Horabik et al. [38], Wang et al. [39], Fu et al. [40] and Chen et al. [41].

Shear box test 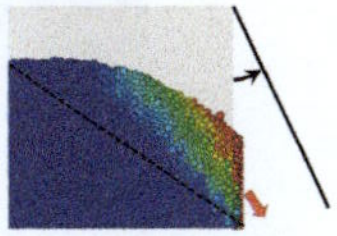	Scale invariant due to the rapid outflow.	The shear angle; the remaining mass.	The shear box is very simple and easy to manufacture, but must have transparent side walls for optical measurements. The tests are simple and quick to execute (simulations and experimentally). Suitable for all sample and PSD sizes.	The remaining mass in the shear box can be determined and used as an additional calibration parameter. For further information see: Derakhshani et al. [36], Do et al. [35], Roessler et al. [14], Lommen et al. [42], Pachon-Morales et al. [43], Mohajeri et al. [25].
Cone test 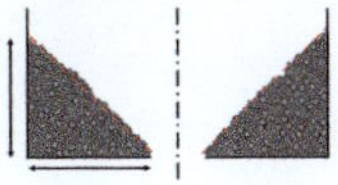	Scale variant due to the variable ratio between the orifice diameter and the particle size.	The shear angle (also called the drained AoR).	The test apparatus is simple and easy to manufacture. Typically, a transparent (glass or poly-glass) cylinder is used. The tests are simple and quick to execute (simulations and experimentally). Typically for small samples and fine-grained materials.	Due to the often cylindrical design it is very difficult to change the diameter of the orifice. The whole slope cannot be observed, only the height against the outer cylinder wall. Thus, a straight slope face is assumed in the calculation of the angle. For further information see: Do et al. [35], Derakhshani et al. [36], Horabik et al. [38] and Fu et al. [40].
Rotating drum test 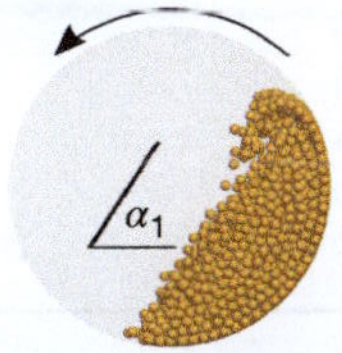	Scale variant.	The dynamic AoR.	The test apparatus is more complex to manufacture. It requires a rotating drum with a transparent side wall and an electrical drive and controller to ensure the drum rotates at constant speed. Compared to the experiments, the simulation model is easier to create. The tests are simple and quick to execute (simulations and experimentally). Typically for smaller samples and fine-grained material (powders).	Recommended to perform this test in the so-called rolling regime which is defined by the Froude number in the range $10^{-4} < Fr < 10^{-2}$ where $Fr = \omega^2 R/g$ with ω the rotation speed, R the drum radius and g the gravitational acceleration. Due to dynamic effects, the differences observed in the AoR from different bulk materials may be small. For further information see: Coetzee [11, 44], Elskamp et al. [37], Rong et al. [45].

(Continued)

Table 1.2 (Continued)

Calibration test	Test properties: Scale effects	Bulk measures for calibration	Test effort, applicability	Comments and further references
Discharge test 	Scale variant due to the variable ratio between the opening and particle size.	The discharge time (duration); the general flow field if coloured particles and a transparent bin are used.	The test apparatus is simple and easy to manufacture. Almost any bin size and shape can be used. The tests are simple and quick to execute (simulations and experimentally). Suitable for all sample and PSD sizes if the opening diameter is selected to accommodate enough of the largest particles.	Stopwatch measurements are inaccurate due to the trickling down of material, and it is often hard to determine the precise moment when the bin is fully emptied. For further information see: Balevicius et al. [46], Derakhshani et al. [36], Hlosta et al. [47], Coetzee and Nel [48], Coetzee [18].
Draw down test 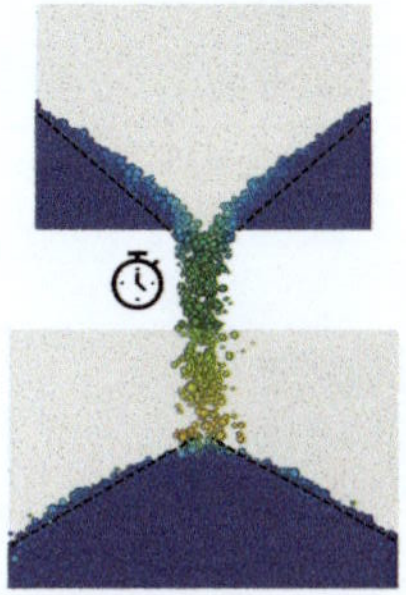	Scale variant.	The shear angle (upper box); the dynamic (poured) AoR (lower box); mass flow rate and/or remaining mass.	The test apparatus is very simple and easy to manufacture, but must have transparent side walls. The tests are simple and quick to execute (simulations and experimentally). Suitable for all small to medium sample and PSD sizes.	This test combines a number of other calibration tests into one: funnel test, shear box test, cone test and the discharge test. Rectangular design for easy change of discharge opening. Different opening settings allow for the studying of arching of cohesive materials. For further information see: Roessler et al. [14], Coetzee [11], Grima and Wypych [49], Derakhshani et al. [36], Do et al. [35], Zhou et al. [50].

Inclined wall friction test 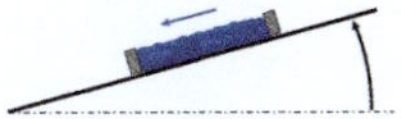	Scale invariant.	Direct measurement of the static friction angle; Measure the sliding or rolling speed to estimate the coefficient of dynamic friction and rolling friction, respectively.	The test setup is more complex to manufacture, if a manual process is not to be followed. The angle of the wall needs to be adjusted using a very slow and smooth action. For this purpose a dedicated motor- or mechanical drive is needed. The tests are simple and quick to execute. Suitable for all particle sizes. Rolling friction can only be estimated for (close to) spherical particle shapes.	Direct measurement of parameter values; hence, no need to replicate the test in a DEM simulation (DMA). For further information see: Barrios et al. [51], Wang et al. [39], Hlosta et al. [47], Agarwal et al. [52].
Direct shear tests 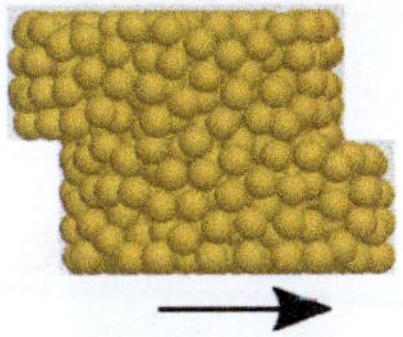	Scale variant.	Angle of internal friction; cohesion (apparent cohesion for non-cohesive materials); dilatant behaviour; material-wall friction angle.	The test apparatus is complex and should be carefully manufactured or, preferably, a commercial unit should be used. Tests are cumbersome to execute and require multiple steps (simulations and experimentally). Limited maximum particle size dependent on the available test cells (experiment). For a given maximum particle size, all PSDs can be accommodated.	To obtain a full yield locus, a series of tests are required which can take long to complete (simulations and experimentally). For detailed procedures, refer to the various standards, such as ASTM D3080/D3080M [53] and ISO 17892-10:2018 [54] for the translational Jenike shear tester, and ASTM D6773-16 [55] for the Schulze ring shear tester (rotational). These standard test procedures require several steps such as pre-shear and shear, especially for cohesive materials. For further information see: Coetzee [11, 18], Mohajeri et al. [25, 57], Simons et al. [56], Bian et al. [58] and Falke et al. [59], Grima and Wypych [60].

(Continued)

Table 1.2 (Continued)

Calibration test	Test properties			Comments and further references
	Scale effects	Bulk measures for calibration	Test effort, applicability	
Confined uniaxial compression test 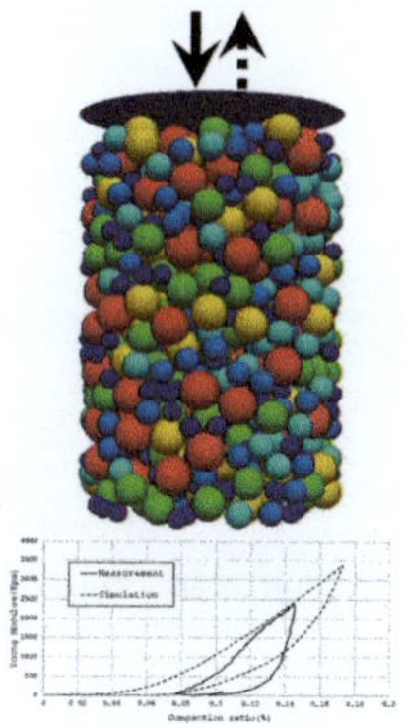	Scale invariant if the container is sufficiently large to accommodate enough particles per cross section.	Bulk stiffness.	Although the cylindrical container is simple and easy to manufacture, the test requires a universal tensile/compression or similar test machine. The simulation is easy to set up. The tests are simple and quick to execute. Suitable for all sample and PSD sizes, by selecting an appropriate container size.	Note that this test should not be confused with the uniaxial compression test [UCT] often used to measure the flow function (unconfined compression strength) of cohesive materials (mainly powders). In the UCT, the sample is first compressed (consolidated) under confined conditions, after which the same sample is compressed under unconfined conditions. If a tensile/compression machine is not available, dead loads can be used to apply the normal load incrementally. This is not ideal, but can provide enough information to estimate bulk stiffness. For further information see: Coetzee [11, 18], Lommen et al. [27], Thakur et al. [61], Donohue et al. [62].
Drop and pendulum tests 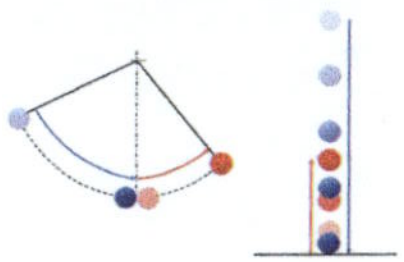	Scale invariant.	Particle–particle coefficient of restitution; particle–wall coefficient of restitution.	The tests are relatively easy and simple to execute, but a means to optically measure the particle motion is needed. Suitable for all particle sizes. Drop tests are not suitable for non-spherical particles, unless very sophisticated equipment and techniques are available to measure the 3D particle trajectory and spin.	Direct measurement of parameter values; hence, no need to replicate the test in a DEM simulation (DMA) For further information see: Barrios et al. [51], Yurata et al. [63], Hastie [64], Hlosta et al. [65], Elskamp et al. [37].

Penetration test 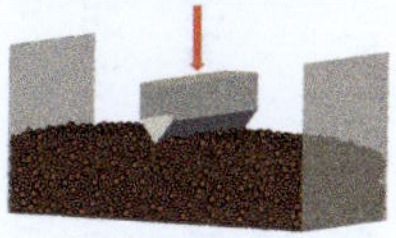	Scale variant due to the ratio of the tip to particle size.	Penetration resistance (resistance of the tip and shaft can be measured independently). If a transparent container is used, and the particles are coloured by layer, the flow field can be observed and measured.	The test apparatus is complex and requires an actuator to push the penetrometer into the material bed at a constant speed. The simulation is easier to set up. The tests are simple and quick to execute. Requires a large sample to cover the container. Suitable for all PSDs.	The container should be large enough to eliminate any wall boundary effects. For further information see: Mohajeri et al. [66, 67], Lommen et al. [42], Coetzee and Els [10].

1.7 Recommended Calibration Procedure for Non-cohesive and Slightly Cohesive Materials

As described earlier, the contact model parameters need to be determined by running a series of DEM simulations with varying parameters. The parameters can be varied systematically or through an optimisation algorithm. Ideally, in both approaches, an automated post-processing of the model results is required since a manual process would not be feasible for a large number of simulations.

The systematic variation of the parameter values over a whole range is time-consuming due to the high number of independent simulations required. However, this approach provides a deep understanding of the general relation between the simulated macroscopic (bulk) behaviour and the (individual) input parameters. Hence, this section focuses on the systematic approach, while Section 1.8 provides an overview of optimisation algorithms.

1.7.1 Ambiguous Parameter Combinations

The ideal situation would be to have an experiment of which the result is dependent on only a single parameter. If such an experiment is available for each of the unknown parameter values, calibration would be straightforward. Unfortunately, most numerical experiments are, however, sensitive to more than one parameter. As an example, take the very common AoR lifting cylinder test. Wensrich and Katterfeld [17] showed that (for non-cohesive materials) the AoR is dependent on both the particle–particle coefficient of sliding friction μ_P and the coefficient of rolling friction μ_r. When such results are plotted, a contour diagram results, which typically looks like the graph presented in Figure 1.3.

Figure 1.3 clearly demonstrates that there are an infinite number of parameter (μ_P and μ_r) combinations which result in the same value of the macroscopic AoR. This example highlights the problem of ambiguous parameter combinations. One calibration test result cannot provide a unique set of two (or more) contact model parameters. Hence, several test results are necessary to find a unique parameter set. The following section describes a procedure which solves the problem of ambiguous parameter combinations for non-cohesive materials.

This approach can also work for 'slightly' cohesive materials. However, with increasing bulk cohesive behaviour, the material flow increases in complexity and the test results are usually less consistent. Therefore, a larger number of tests are required to decrease the parameter range. To illustrate this point, a comparison of the AoR test with and without cohesion is shown in Figure 1.4. The calibration of cohesive materials is described in more detail in Section 1.7.

1.7.2 Solving for an Unambiguous Parameter Combination

The problem of ambiguous parameter combinations is equivalent to simultaneously solving a set of mathematical equations with a number of unknown variables or (calibration) parameters. A unique solution can only be obtained if the number of

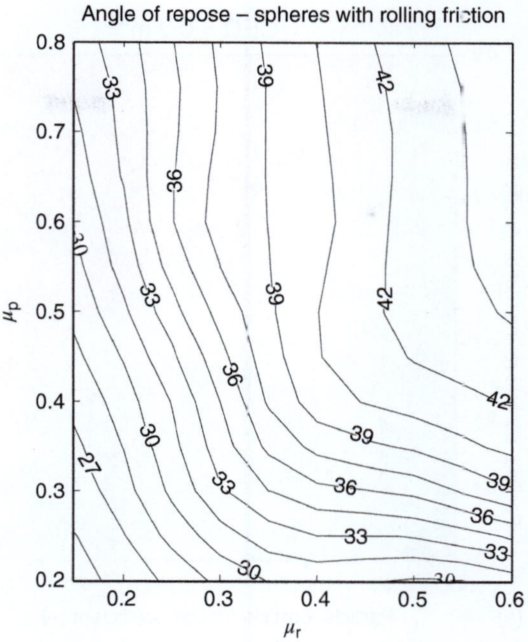

Figure 1.3 AoR simulation results according to Wensrich and Katterfeld et al. [17] using a lifting cylinder test for a systematic variation of particle–particle sliding friction μ_P and rolling friction μ_r. A constant macroscopic AoR is represented by an isoline. No cohesion was considered.

independent equations is equal to the number of variables. Using this analogy, the calibration test(s) should produce the same number (or more) of 'independent' test results as the number of DEM parameters to which the model is sensitive. The explanation in Section 1.4 and the summary in Table 1.1 help to identify the most influential parameter(s) for a given bulk property.

This analogy clearly explains why calibration tests which produce more than one test result (or bulk measure) are preferred. For example, the draw down test and the direct shear test produce four results each. However, not all of the test results can be called independent. For instance, the two AoR results and the remaining mass in the upper or lower chamber of the draw down test are dependent on each other. The higher the AoR in the upper chamber, the higher is the remaining mass in that chamber, and vice versa. In the direct shear test, the shear response at different normal stresses can be used as well as the internal friction angle and the time to reach failure.

For each bulk measure (either from a single test or multiple tests), a diagram such as the one presented in Figure 1.3 can be produced. The diagrams can then be superimposed to obtain a unique or unambiguous set of parameter values, i.e. a single set of parameter values which satisfies all test results. However, there is no guarantee that, in all cases, the isolines (one for each bulk measure) will intersect at a single point as described in Derakhshani et al. [36]. Often the measurement error needs to be considered to ensure a significant overlap of the test results as shown in detail by Roessler et al. [14].

Figure 1.5 shows how the different draw down test results (shear angle in the upper chamber, AoR in the lower chamber, mass in the lower chamber, and mass

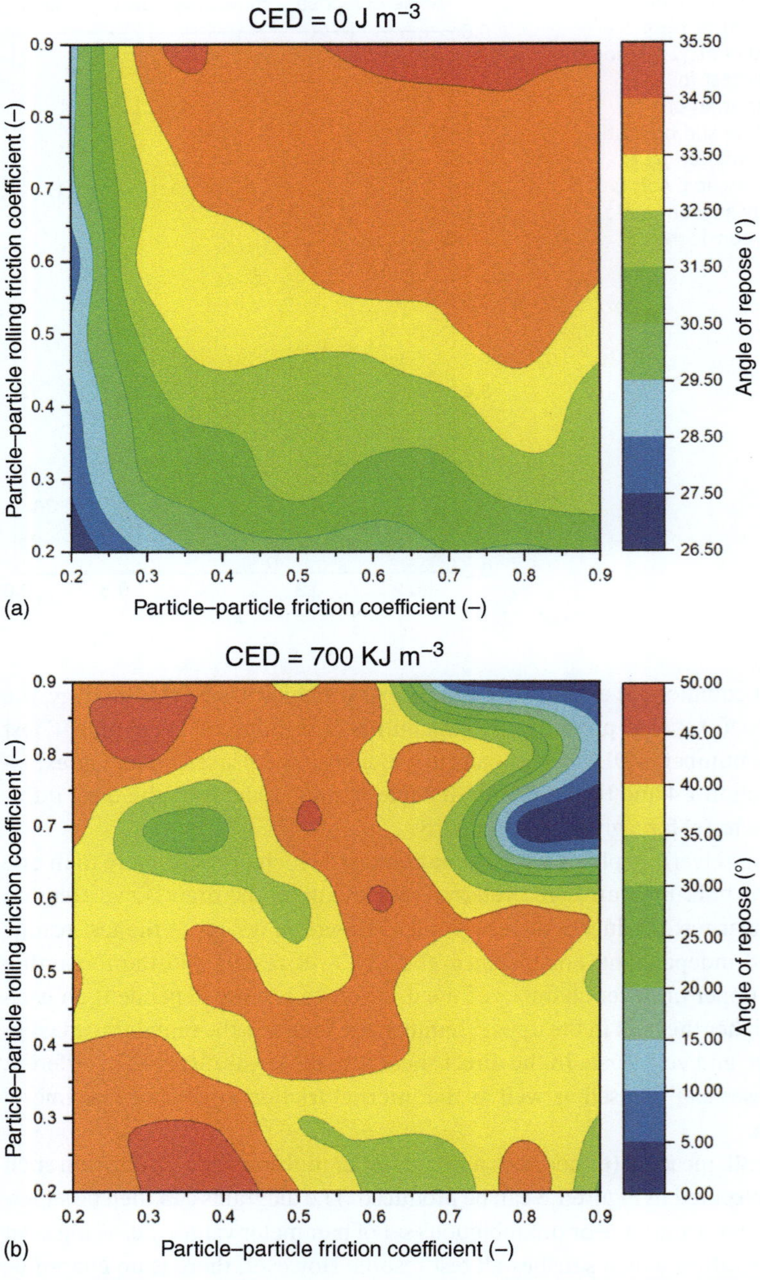

Figure 1.4 Comparison of AoR simulation results with varying particle–particle (sliding) friction and rolling friction coefficient for non-cohesive (a) and cohesive materials (b). Similar to Pachon-Morales et al. [43] for non-cohesive and slightly cohesive materials, a direct relation between the frictional values and the AoR can be shown. This is not the case for the cohesive material where a more irregular and complex behaviour can be seen.

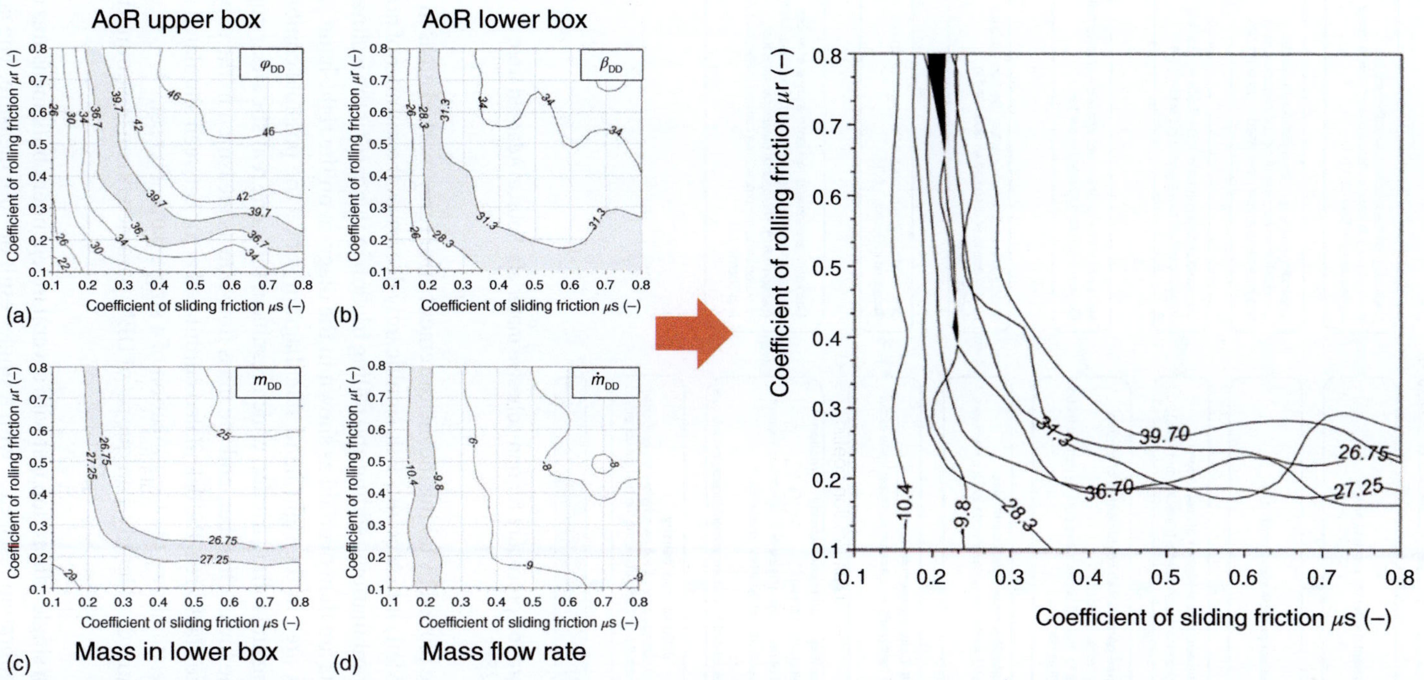

Figure 1.5 Combination of draw down test results for gravel with consideration of the measurement error for the identification of a unique parameter set. Source: Adapted from Roessler et al. [14].

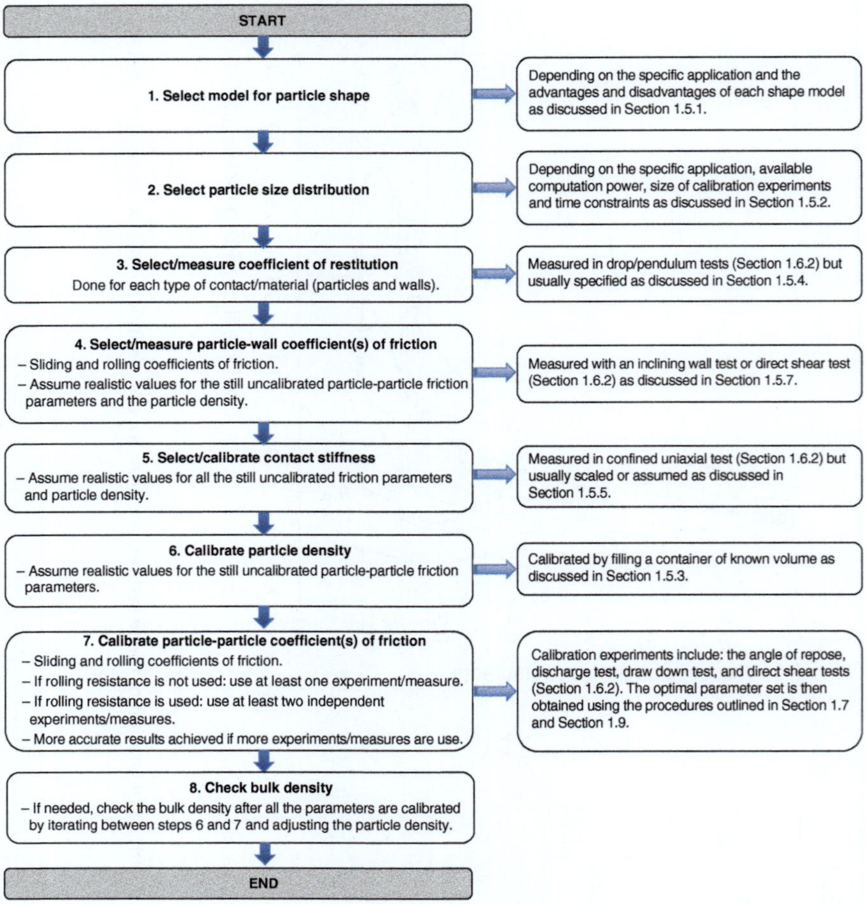

Figure 1.6 Calibration procedure for non-cohesive materials. Source: Adapted from Katterfeld et al. [28].

flow rate) can be plotted by taking the measurement error (grey marked areas in the diagrams on the left) into account. This significantly decreases the number of possible parameter combinations (sliding and rolling friction in this case) if the isolines of the contour plots are then overlaid as shown in the diagram on the right-hand side. The small black area in the right-hand side diagram shows the possible combination of the sliding and rolling friction which results in the same realistic shear angle, AoR, mass in the lower chamber, and the mass flow rate as measured in the experiment. Within the black area, the calculated cumulated error between simulated and measured test result allows the identification of a single parameter set.

The steps required to identify a unique set of DEM parameters can be summarised as follows:

1. Make use of a single calibration test with several independent bulk measures or a number of calibration tests, each with a single result or bulk measure. The latter case requires more effort in terms of experiments and simulations.

2. Determine the measurement errors for each test result (bulk measure) and consider them in the analysis of the DEM results.
3. Identify the parameter ranges which provide realistic simulation results for each test.
4. Calculate the cumulative error between simulation results and experimental measurements and identify the parameter set with a minimum error.

In the given examples of the lifting cylinder test (Figure 1.3) and the draw down test (Figure 1.5), only the influence of the particle–particle coefficients of sliding and rolling friction was shown. Although these two parameters are the ones which have the most significant influence on the modelled bulk behaviour of non-cohesive materials, the other parameters should also be carefully considered. This leads to the obvious question: is it necessary to vary all these parameter values systematically? This would require a tremendous number of independent simulation runs, which is not feasible. However, it has been shown that if the calibration procedure of individual parameters is executed in a specific sequence, the need for iterations and finding the unique (optimal) set of parameters can be minimised.

Based on the experience of a large number of academic and industrial experts, the calibration procedure as shown in the flowchart in Figure 1.6 was developed for non-cohesive materials, as presented in Katterfeld et al. [28]. The flowchart includes references to the sections where the influence of the specific parameter on the bulk behaviour is explained as well as the proposed experiment(s) to calibrate each parameter.

1.8 Outlook on the Calibration of Cohesive Materials

Cohesion, usually through the effects of moisture, can significantly alter the bulk behaviour of granular materials. However, the DEM research and modelling community do not yet have the same level of confidence in any DEM model to accurately predict the bulk behaviour of cohesive materials compared to that of non-cohesive materials. The exact reason for this is not known, and it might be that we are not yet able to successfully calibrate the cohesive parameters, and/or the cohesive contact models that are currently available do not capture the real physics accurately enough. For completeness, a short summary of the published attempts made to calibrate cohesive materials is presented here.

Grima [68] and Grima and Wypych [49] developed and validated a strategy for calibrating the parameters of dry and wet black coal and bauxite. First, the material was calibrated in a dry non-cohesive state based on flow property measurements and bench scale tests and using the Hertz–Mindlin (no slip) contact model. Thereafter, the material was tested in a moist cohesive state and the appropriate cohesive parameters introduced to the DEM contact model. The cohesive parameters were calibrated, while the non-cohesive parameter values remained unchanged. Two experimental tests were used to calibrate the parameters. First, a swing-arm slump test with a split cylinder was developed to measure the AoR. The second test was based on a cylindrical draw down test, where the shear angle (upper box), poured AoR (lower box), and the mass flow rate (estimated from high-speed photography) were measured. In addition, the impact force of the material onto the bottom plate was measured.

Due to the cohesive effects, the shape of the formed piles was irregular, which made it difficult to define the AoR in the slump test. A quantitative approach was then followed, and the general shape of the heap used for comparison. In the draw down test, the cohesive material slumped down onto the lower plate, and the cohesive parameter (the cohesive energy density) had no significant effect on the poured AoR. However, it had a significant effect on the shear angle, and in the end, this was the only measured parameter from the draw down test that could be used for calibration purposes. However, a single value of the cohesion energy density, which accurately predicted the slump test AoR and the draw down shear angle, could not be found. In the end, the study concluded that a calibrated cohesion parameter for one system might not be suitable for another system.

Katterfeld et al. [69] used two different cohesive contact models and the draw down test to calibrate the parameters for wet gypsum. Both models under-predicted the shear angle in the upper box of the draw down test and over-predicted the poured AoR in the lower box.

Roessler and Katterfeld [29] made use of the lifting cylinder test to calibrate the cohesive parameter for wet sand. The AoR showed a very high level of variability, ranging from 40.4° to 84.3° for the same material in repeated tests. Hence, the macroscopic flow behaviour was rather analysed where three distinct and reproducible phases could be identified during the upward motion of the cylinder and used as calibration criteria. This study, however, made use of a single moisture content, and this approach still needs to be verified for different moisture contents and different bulk materials.

Ajmal et al. [70], on the other hand, showed that the draw down test could be used to successfully calibrate sliding friction, rolling friction, and cohesion parameter for wet sand (10% moisture content). It was shown that the blockage or arching of material at the opening was independent of sliding and rolling friction and depended only on the cohesion parameter. Comparing these results to experimental observations for various opening sizes, a narrow band of values for the cohesion parameter could be established. A unique set of parameter values could then be obtained by comparing the AoR, formed in the lower box, and the total mass that flowed to the bottom box, to experimental measurement. It was noted that the shear angle, formed in the upper box, could not be used since it was not reproducible.

Mohajeri et al. [57] used the ring shear test to calibrate the parameters of cohesive (moist) coal by using an elasto-plastic cohesion contact model. This work was further extended [71] in a study where a combination of calibration experiments was used to calibrate the parameters of cohesive iron ore. The tests included a ring shear test, a shear box (ledge AoR), and a penetration test.

Carr et al. [72] investigated the use of two contact models to simulate the behaviour of cohesive iron ore with varying moisture content. The material was allowed to flow from a feeding conveyor onto an impact plate, where the build-up of material was observed. It was concluded that only the liquid-bridge model was capable of simulating the clumping effect of the material which was observed during the experiments. In a further study [73], a combination of the shear box (ledge test), draw down, and a conveyor impact plate test was used. By varying the cohesive parameter, the draw down test was used to investigate flowing and arching cases. The irregular slopes that formed made it difficult to measure the angles, and it was concluded that in the shear box and the draw down test, the residual mass was a more definite and reproducible calibration measure.

1.9 Optimisation Approaches Applied to the Calibration Process

An increasing number of calibration parameters increases the problem of ambiguous parameter sets and brings classical visualisation techniques such as the contour plots, shown in Section 1.6, to their limits. Hence, if more than two parameters should be calibrated, for example, if one would include cohesion or particle upscaling, the use of optimisation algorithms might be mandatory. Also, an increasing number of calibration test results such as an AoR, shear angle, and mass flow rate from a draw down test or different yield loci from a direct shear test lead to complex analyses with a lot of processing steps. The aim of the calibration process is to find a set of DEM parameter values in such a way that the simulation results of the physical calibration experiments are the most accurate (optimal) as compared to the experimental results. Hence, the task of the optimisation process is to minimise the difference between the experimental and simulation results.

It was already discussed that multiple calibration test results are necessary for finding an unambiguous parameter set. The calibration test results are the so-called objective function values or target values for the optimisation algorithm. Hence, only such algorithms can be used which can handle multiple target values. However, besides the problem of handling multiple parameters and multiple target criteria, the biggest advantage of optimisation algorithms is the reduced number of calibration simulations. This addresses one of the largest bottlenecks in the whole calibration process, namely the time required to run a whole series of calibration simulations.

Different optimisation schemes are available, and Richter et al. [74] identified the following characteristics of typical DEM results as significant to consider in the selection process:

- nonlinear objective function topology,
- expensive computation,
- gradients cannot be obtained directly,
- limitations of the search space in the form of parameter boundaries (restricted optimization problem),
- there can be more than one optimum or many local optima,
- rough objective function surface due to stochastic effects and numerical noise,
- more than one objective function value is necessary for calibration,
- discontinuities can occur (e.g. bridging in funnel flow experiments due to high friction values).

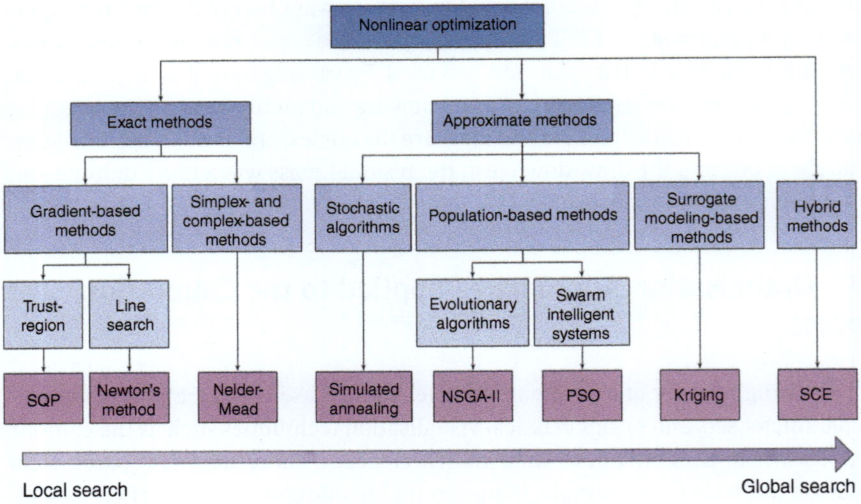

Figure 1.7 Overview of optimisation algorithm. SQP = sequential quadratic programming, NSGA = non-dominated sorting genetic algorithm, PSO = particle swarm optimization, SCE = shuffled complex evolution. Source: Richter et al. [74]/with permission of Elsevier.

Richter et al. [74] also provides a comprehensive overview of the different optimisation approaches which are generally known (Figure 1.7). A systematic review of these approaches is essential since different terms are used in literature for the same optimisation approach. For example, Hess et al. [75] used the term particle swarm optimisation whereas Do et al. [35] used the term genetic algorithms.

An increasing number of publications explain the use of 'Artificial Intelligence' or 'Machine Learning' to optimise the calibration process. However, often the actual algorithm used is similar to those from publications which make use of different terms.

For the DEM application, the use of surrogate models can be identified as one of the best optimisation approaches. Many DEM publications focus on this approach. Most surrogate models are capable of identifying a global optimum on a rough objective function surface with many local optima within a few iterations. Further, they can consider parameter limitations and multi-objective problems and discontinuities.

The surrogate model approach generates, in the first step, a virtual approximation of the objective functions. Only a few calibration simulations are necessary for this. The sampling method for selecting the initial parameters is an important step. Based on the generated surrogate model, a set of optimised parameters is predicted and evaluated by the so-called acquisition function. Then, an additional calibration simulation is undertaken, and the difference to the target values is calculated. If the difference is larger than stopping criteria, the surrogate model is refined by the new simulation results and the whole process is repeated. Richter et al. [74] have suggested a general description of the necessary steps to use the surrogate model approach. The flow diagram in Figure 1.8 summarises the flow of the so-called Generalised Surrogate Modelling-based Calibration (GSMC).

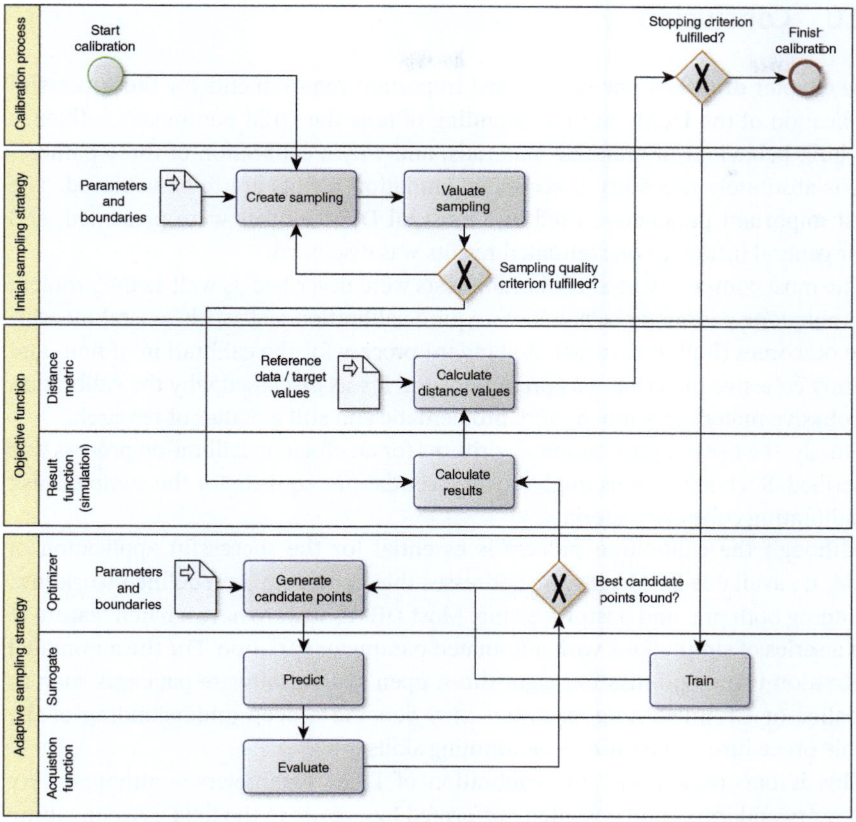

Figure 1.8 Component structure and process of generalised surrogate modelling-based calibration (GSMC). Source: Richter et al. [74]/with permission of Elsevier.

To calculate and predict the surrogate model, several approaches are known. They can be separated in probabilistic and non-probabilistic approaches. Probabilistic surrogate models include the calculation of the probability for the prediction of the estimated value, which provides advantages in the evaluation process. Richter et al. [74] have analysed a number of different surrogate algorithms, which include Gaussian process regression (GPR), artificial neural networks, multi-adaptive regression splines, and universal Kriging. Rackl and Hanley [76] and Kriging and Benvenuti et al. [77] used artificial neural networks. Richter et al. [74] could achieve the best results with the GPR which belongs to the probabilistic approaches. Due to the probabilistic surrogate model and the probabilistic acquisition function, the approach can be classified as a Bayesian optimisation approach. This approach and the Gaussian model were also used by Hartmann et al. [78].

Since Richter et al. [74] and Roessler et al. [14] used the same calibration tests and data, it could be shown that the use of the optimisation algorithm resulted in the same final DEM parameter set as the systematic parameter selection process where a full factorial design was used. While the systematic selection process required at least 64 simulation runs, the GSMC with GPR needed less than 23 simulation runs.

1.10 Conclusion

This chapter discusses one of the most important requirements for the successful application of the DEM: an understanding of how the DEM parameters influence the bulk behaviour of granular materials, and why a calibration of these parameters is absolutely necessary if accurate simulation results are to be achieved. The most important parameters used in almost all DEM models were presented, and their general influence on simulated results was discussed.

The most commonly used calibration tests were described as well as the problem of ambiguous parameter sets, which require calibration tests with several measurable outcomes (bulk measures). A standard process for the calibration of non- and slightly cohesive materials was presented, and it was discussed why the calibration of cohesive materials is much more problematic and still a matter of research.

Finally, the use of optimisation algorithms for an efficient calibration process was described. Such algorithms might be an essential prerequisite for the complex task of calibrating cohesive materials.

Although the calibration process is essential for the successful application of DEM, no available DEM software addresses this issue with a predefined workflow, including both pre- and postprocessing. Most DEM software have a batch feature to run a series of simulations with automated parameter variation. For the automated calibration using optimisation algorithms, open source software packages such as Decalioc or GrainLearning exist, but they demand a deep understanding of the whole procedure and require programming skills.

This is one reason why the calibration of DEM parameters – although very necessary and important – is often performed by experts in the field as a consulting service. Such services must include suitable experimental tests and the run and analysis of a series of calibration simulations. Experience proves that, with the current state of the art in DEM calibration, knowledge of the physical behaviour of bulk materials in close combination with simulation know-how is required to obtain accurate calibration results. However, metamodel-based approaches like those described by Richter and Will [79] might lead to an improved understanding of the DEM parameters and their influence on the macroscopic behaviour of bulk materials and, in the future, significantly reduce the complexity of the calibration process.

Glossary

Angle of repose
Bulk calibration approach
Bulk cohesion
Bulk density
Bulk friction
Bulk stiffness
Calibration philosophies
Calibration process
Calibration test

Coarse graining
Coefficient of restitution
Cohesive material
Combined scaling
Contact damping
Contact model
Contact properties
Contact stiffness
DEM parameters
Direct measurement approach
Dissipation of energy
Draw down test
Exact scaling
Hybrid scaling
Lifting cylinder
Multi-sphere particle
Non-cohesive material
Optimisation
Parameter combination
Parameter set
Particle density
Particle shape
Particle size distribution
Physical bulk properties
Porosity
Rolling friction
Scalping
Sliding friction

References

1 Coetzee, C.J., Basson, A.H., and Vermeer, P.A. (2007). Discrete and continuum modelling of excavator bucket filling. *Journal of Terramechanics* 44: 177–186.

2 Cundall, P.A. and Strack, O.D. (1979). Discrete numerical model for granular assemblies. *Géotechnique* 29 (1): 47–65.

3 O'Sullivan, C. (2011). *Particulate discrete element modelling, a geomechanics perspective, Applied Geotechnics*, vol. 4. London and New York: Spon Press ISBN13: 978-0-203-88098-2.

4 Thornton, C. (2015). *Granular dynamics, contact mechanics and particle system simulation, A DEM study*. Cham: Springer ISBN 978-3-319-18710-5.

5 Hazzar, L., Nuth, M., and Chekired, M. (2020). DEM simulation of drained triaxial tests for glass-beads. *Powder Technology* 364: 123–134.

6 Mesnier, A., Peczalski, R., Mollon, G., and Vessot-Crastes, S. (2020). Mixing of bi-dispersed milli-beads in a rotary drum: mechanical segregation analyzed by lab-scale experiments and DEM simulation. *Processes* 8 (9): 1166.

7 Coetzee, C.J. (2017). Review: calibration of the discrete element method. *Powder Technology* 310: 104–142.

8 Mitarai, N. and Nori, F. (2006). Wet granular materials. *Advances in Physics* 55 (1, 2): 1–45.

9 Seville, J.P.K., Willet, C.D., and Knight, P.C. (2000). Interparticle forces in fluidisation: a review. *Powder Technology* 113: 261–268.

10 Coetzee, C.J. and Els, D.N.J. (2009). Calibration of discrete element parameters and the modelling of silo discharge and bucket filling. *Computers and Electronics in Agriculture* 65: 198–212.

11 Coetzee, C.J. (2020). Calibration of the discrete element method: strategies for spherical and non-spherical particles. *Powder Technology* 364: 851–878.

12 Guo, Y., Zhao, C., Markine, V. et al. (2020). Discrete element modelling of railway ballast performance considering particle shape and rolling resistance. *Railway Engineering Science* 28: 382–407.

13 Markauskas, D., Ramírez-Gómez, A., Kacianauskas, R., and Zdancevicius, E. (2015). Maize grain shape approaches for DEM modelling. *Computers and Electronics in Agriculture* 118: 247–258.

14 Roessler, T., Richter, C., and Katterfeld, A. (2019b). Development of a standard calibration procedure for the DEM parameters of cohesionless bulk materials. Part I - solving the problem of ambiguous parameter combinations. *Powder Technology* 343: 803–812.

15 Wensrich, C.M., Katterfeld, A., and Sugo, D. (2014). Characterisation of the effects of particle shape using a normalised contact eccentricity. *Granular Matter* 16 (3): 327–337.

16 Ai, J., Chen, J.F., Rotter, J.M., and Ooi, J.Y. (2011). Assessment of rolling resistance models in discrete element simulations. *Powder Technology* 206: 269–282.

17 Wensrich, C. and Katterfeld, A. (2012). Rolling friction as a technique for modelling particle shape in DEM. *Powder Technology* 217: 409–417.

18 Coetzee, C.J. (2016). Calibration of the discrete element method and the effect of particle shape. *Powder Technology* 297: 50–70.

19 Stahl, M. and Konietzky, H. (2011). Discrete element simulation of ballast and gravel under special consideration of grain-shape, grain-size and relative density. *Granular Matter* 13: 417–428.

20 Chen, W., Donohue, T., Katterfeld, A., and Williams, K. (2017). Comparative discrete element modelling of a vibratory sieving process with spherical and rounded polyhedron particles. *Granular Matter* 19 (81) 1–12.

21 Alonso-Marroquín, F., Ramírez-Gómez, A., González-Montellano, C. et al. (2013). Experimental and numerical determination of mechanical properties of polygonal wood particles and their flow analysis in silos. *Granular Matter* 15: 811–826.

22 Lenaerts, B., Aertsen, T., Tijskens, E. et al. (2014). Simulation of grain–straw separation by discrete element modeling with bendable straw particles. *Computers and Electronics in Agriculture* 101: 24–33.

23 Thakur, S.C., Ooi, J.Y., and Ahmadian, H. (2016). Scaling of discrete element model parameters for cohesionless and cohesive solid. *Powder Technology* 293: 130–137.

24 Roessler, T. and Katterfeld, A. (2018). Scaling of the angle of repose test and its influence on the calibration of DEM parameters using upscaled particles. *Powder Technology* 330: 58–66.

25 Mohajeri, M.J., Helmons, R.L., van Rhee, C., and Schott, D.L. (2020a). A hybrid particle-geometric scaling approach for elasto-plastic adhesive DEM contact models. *Powder Technology* 369: 72–87.

26 Katterfeld, A., Donohue, T., and Ilic, D. (2012). Application of the discrete element method in mechanical conveying of bulk materials. In: *07th International Conference for Conveying and Handling of Particulate Solids*, 10–13 September 2012, Friedrichshafen, Germany.

27 Lommen, S., Schott, D.L., and Lodewijks, G. (2014). DEM speedup: stiffness effects on behaviour of bulk material. *Particuology* 12: 107–112.

28 Katterfeld, A., Coetzee, C., Donohue, T., et al. (2019). White Paper: Calibration of DEM parameters for cohesionless bulk materials under rapid flow conditions and low consolidation. https://www.researchgate.net/publication/334129159_Calibration_of_DEM_Parameters_for_Cohesionless_Bulk_Materials_under_Rapid_Flow_Conditions_and_Low_Consolidation.

29 Roessler, T. and Katterfeld, A. (2019a). DEM parameter calibration of cohesive bulk materials using a simple angle of repose test. *Particuology* 45: 105–115.

30 Tan, Y., Yu, Y., Fottner, J., and Kessler, S. (2021). Automated measurement of the numerical angle of repose (aMAoR) of biomass particles in EDEM with a novel algorithm. *Powder Technology* 388: 462–473.

31 Muller, D., Fimbinger, E., and Brand, C. (2021). Algorithm for the determination of the angle of repose in bulk material analysis. *Powder Technology* 373: 598–605.

32 Klanfar, M., Korman, T., Domitrovic, D., and Herceg, V. (2021). Testing the novel method for angle of repose measurement based on area-weighted average slope of a triangular mesh. *Powder Technology* 387: 396–405.

33 Qi, L., Chen, Y., and Sadek, M. (2019). Simulations of soil flow properties using the discrete element method (DEM). *Computers and Electronics in Agriculture* 157: 254–260.

34 Zhou, L., Yu, J., Liang, L. et al. (2021). DEM parameter calibration of maize seeds and the effect of rolling friction. *Processes* 9 (914): https://doi.org/10.3390/pr9060914.

35 Do, H.Q., Aragon, A.M., and Schott, D.L. (2018). A calibration framework for discrete element model parameters using genetic algorithms. *Advanced Powder Technology* 29: 1393–1403.

36 Derakhshani, S.M., Schott, D.L., and Lodewijks, G. (2015). Micro macro properties of quartz sand: experimental investigation and DEM simulation. *Powder Technology* 269: 127–138.

37 Elskamp, F., Kruggel-Emden, H., Henning, M., and Teipel, U. (2017). A strategy to determine DEM parameters for spherical and non-spherical particles. *Granular Matter* 19 (46): 1–13.

38 Horabik, J., Parafinuik, P., and Molenda, M. (2017). Discrete element modelling study of force distribution in a 3D pile of spherical particles. *Powder Technology* 312: 194–203.

39 Wang, L., Li, R., Wu, B. et al. (2018). Determination of the coefficient of rolling friction of an irregularly shaped maize particle group using physical experiment and simulations. *Particuology* 38: 185–195.

40 Fu, J.-J., Chen, C., Ferellec, J.-F., and Yang, J. (2020). Effect of particle shape on repose angle based on hopper flow test and discrete element method. *Advances in Civil Engineering* 2020: 8811063. https://doi.org/10.1155/2020/8811063.

41 Chen, Z., Wassgren, C., Veikle, E., and Ambrose, K. (2020). Determination of material and interaction properties of maize and wheat kernels for DEM simulation. *Biosystems Engineering* 195: 208–226.

42 Lommen, S., Mohajeri, M., Lodewijks, G., and Schott, D. (2019). DEM particle upscaling for large-scale bulk handling equipment and material interaction. *Powder Technology* 352: 273–282.

43 Pachon-Morales, J., Do, H., Colin, J. et al. (2019). DEM modelling for flow of cohesive lignocellulosic biomass powders: model calibration using bulk tests. *Advanced Powder Technology* 30: 732–450.

44 Coetzee, C.J. (2019). Particle upscaling: calibration and validation of the discrete element method. *Powder Technology* 344: 487–503.

45 Rong, W., Feng, Y., Schwartz, P. et al. (2020). Sensitivity analysis of particle contact parameters for DEM simulation in a rotating drum using response surface methodology. *Powder Technology* 362: 604–614.

46 Balevicius, R., Kacianauskas, R., Mroz, Z., and Sielamowicz, I. (2006). Discrete element method applied to multiobjective optimization of discharge flow parameters in hoppers. *Structural and Multidisciplinary Optimization* 31: 163–175.

47 Hlosta, J., Jezerská, L., Rozbroj, J. et al. (2020). DEM investigation of the influence of particulate properties and operating conditions on the mixing process in rotary drums: part 1 – determination of the DEM parameters and calibration process. *Processes* 8 (222): https://doi.org/10.3390/pr8020222.

48 Coetzee, C. and Nel, R. (2014). Calibration of discrete element properties and the modelling of packed rock beds. *Powder Technology* 264: 332–342.

49 Grima, A.P. and Wypych, P.W. (2011). Development and validation of calibration methods for discrete element modelling. *Granular Matter* 13: 127–132.

50 Zhou, Y.C., Xu, B.H., Yu, A.B., and Zulli, P. (2002). An experimental and numerical study of the angle of repose of coarse spheres. *Powder Technology* 125: 45–54.

51 Barrios, G.K.P., de Carvalho, R.M., Kwade, A., and Tavares, L.M. (2013). Contact parameter estimation for DEM simulation of iron ore pellet handling Gabriel. *Powder Technology* 248: 84–93.

52 Agarwal, A., Tripathi, A., Tripathi, A. et al. (2021). Rolling friction measurement of slightly non-spherical particles using direct experiments and image analysis. *Granular Matter* 23 (60): 1–14.

53 ASTM D3080/D3080M (2011). *Standard test method for direct shear test of soils under consolidated drained conditions.* West Conshohocken, PA: ASTM International.

54 ISO 17892-10:2018(2018). Geotechnical investigation and testing — Laboratory testing of soil — Part 10: Direct shear tests.

55 ASTM D6773-16 (2016). *Standard test method for bulk solids using schulze ring shear tester.*

56 Simons, T.A.H., Weiler, R., Strege, S. et al. (2015). A ring shear tester as calibration experiment for DEM simulations in agitated mixers – a sensitivity study. *Procedia Engineering* 102: 741–748.

57 Mohajeri, M.J., Do, H.Q., and Schott, D.L. (2020b). DEM calibration of cohesive material in the ring shear test by applying a genetic algorithm framework. *Advanced Powder Technology* 31 (5): 1838–1850.

58 Bian, X., Li, W., Qian, Y., and Tutumluer, E. (2019). Micromechanical particle interactions in railway ballast through DEM simulations of direct shear tests. *International Journal of Geomechanics* 19 (5): 1–19.

59 Falke, T., de Payrebrune, K.M., Kirchhof, S. et al. (2018). An alternative DEM parameter identification procedure based on experimental investigation: a case study of a ring shear cell. *Powder Technology* 328: 227–234.

60 Grima, A. and Wypych, P. (2013). Effect of particle properties on the discrete element simulation of wall friction. In: *ICBMH 2013 - 11th International Conference on Bulk Materials Storage, Handling and Transportation.*

61 Thakur, S.C., Ahmadian, H., Sun, J., and Ooi, J.Y. (2014). An experimental and numerical study of packing, compression, and caking behaviour of detergent powders. *Particuology* 12: 2–12.

62 Donohue, T.J., Wensrich, C.M., and Reid, S. (2016). On the use of the uniaxial shear test for DEM calibration. In: *The 7th International Conference on Discrete Element Methods*. China: Dalian (1–4 August 2016).

63 Yurata, T., Gidaspow, D., Piumsomboon, P., and Chalermsinsuwan, B. (2021). The importance of parameter-dependent coefficient of restitution in discrete element method simulations. *Advanced Powder Technology* 32: 1004–1012.

64 Hastie, D.B. (2013). Experimental measurement of the coefficient of restitution of irregular shaped particles impacting on horizontal surfaces. *Chemical Engineering Science* 101: 828–836.

65 Hlosta, J., Zurovec, D., Rozbroj, J. et al. (2018). Experimental determination of particle–particle restitution coefficient via double pendulum method. *Chemical Engineering Research and Design* 135: 222–233.

66 Mohajeri, M., van Rhee, C., and Schott, D.L. (2018). Penetration resistance of cohesive iron ore: a DEM study. In: *9th International Conference on Conveying and Handling of Particulate Solids, CHoPS 2018.*

67 Mohajeri, M.J., de Kluijver, W., Helmons, R.L.J. et al. (2021a). A validated co-simulation of grab and moist iron ore cargo: replicating the cohesive and stress-history dependent behaviour of bulk solids. *Advanced Powder Technology* 32 (4): 1157–1169.

68 Grima, A.P. (2011). Quantifying and modelling mechanisms of flow in cohesionless and cohesive granular materials. PhD thesis. University of Wollongong.

69 Katterfeld, A., Roessler, T., and Chen, W. (2018). Calibration of the DEM parameters of cohesive bulk materials for the optimisation of transfer chutes. In: *CHoPS 2018, 9th International Conference on Conveying and Handling of Particulate Solids*. (10th–14th September 2018).

70 Ajmal, M., Roessler, T., Richter, C., and Katterfeld, A. (2020). Calibration of cohesive DEM parameters under rapid flow conditions and low consolidation stresses. *Powder Technology* 374: 22–32.

71 Mohajeri, M.J., van Rhee, C., and Schott, D.L. (2021b). Replicating cohesive and stress-history-dependent behavior of bulk solids: feasibility and definiteness in DEM calibration procedure. *Advanced Powder Technology* 32: 1532–1548.

72 Carr, M.J., Chen, W., Williams, K., and Katterfeld, A. (2016). Comparative investigation on modelling wet and sticky material behaviours with a simplified JKR cohesion model and liquid bridging cohesion model in DEM. In: *ICBMH 2016 - 12th International Conference on Bulk Materials Storage, Handling and Transportation, Proceedings.*

73 Carr, M.J., Roessler, T., Otto, H. et al. (2019). Calibration procedure of discrete element method (DEM) parameters for cohesive bulk materials. In: *13th International Conference on Bulk Materials Storage, Handling & Transportation.*

74 Richter, C., Roessler, T., Kunze, G. et al. (2019). Development of a standard calibration procedure for the DEM parameters of cohesionless bulk materials – part II: efficient optimization-based calibration. *Powder Technology* 360: 967–976. https://doi.org/10.1016/j.powtec.2019.10.052.

75 Hess, G., Richter, C., and Katterfeld, A. (2016). Simulation of the dynamic interaction between bulk material and heavy equipment - calibration and validation. In: *ICBMH 2016: 12th International Conference on Bulk Materials Storage, Handling and Transportation; Proceedings, 11–14 July 2016* (ed. D. Hastie), 427–436. Darwin, Australia: The Institution of Engineers.

76 Rackl, M. and Hanley, K.J. (2016). A methodical calibration procedure for discrete element models. *Powder Technology* 307: 73–83. https://doi.org/10.1016/j.powtec.2016.11.048.

77 Benvenuti, L., Kloss, C., and Pirker, S. (2016). Identification of DEM simulation parameters by Artificial Neural Networks and bulk experiments. *Powder Technology* 291: 456–465. https://doi.org/10.1016/j.powtec.2016.01.003.

78 Hartmann, P., Cheng, H., and Thoeni, K. (2021). Performance study of iterative Bayesian filtering to develop an efficient calibration framework for DEM. *Computers and Geotechnics* 141: https://doi.org/10.1016/j.compgeo.2021.104491.

79 Richter, C. and Will, F. (2021). Introducing metamodel-based global calibration of material-specific simulation parameters for discrete element method. *Minerals* 11: 848. https://doi.org/10.3390/min11080848.

2

Simulation of Transfer Chutes

Dusan Ilic[1] and André Katterfeld[2]

[1] *University of Newcastle, School of Engineering, Newcastle Institute for Energy and Resources, University Drive, Callaghan, NSW 2308, Australia*
[2] *Otto von Guericke University Magdeburg, Chair of Material Handling, Institute of Logistics and Material Handling Systems, Universitätsplatz 2, 39106 Magdeburg, Germany*

2.1 Introduction

Transfer chutes play a major role in almost all bulk solids handling operations, including feeding, transfer, loading, and discharge. In these applications, efficient, trouble-free performance is required. Wear needs to be controlled, flow blockage and spillage should be prevented, and maintenance costs should be minimised. A few transfer chute examples are shown in Figure 2.1. To be considered fit-for-purpose, transfer chutes demand adequate attention from the preliminary stages of engineering design. Unfortunately, their design is often overlooked, conflicted by the relative importance of other front-end criteria (e.g. conveyor, structural, maintenance, health, and safety considerations) and a general lack of understanding of the interrelationships between chute functionality and conveyor system handling performance with respect to flow, build-up, blockage, and wear.

Premature downtime and bottlenecks due to ill-considered design aspects can severely impact business. In high-throughput shipping port terminals, for example, minutes of unplanned operational downtime can lead to a reduction in export capacity of thousands of tonnes, translating to significant lost revenue that can seldom be recouped within the financial year. Conversely, modification of chutes in unplanned brownfield situations with inadequate attention to key design aspects and low investment can actually lead to ongoing repairs and maintenance.

While the design of transfer chutes varies widely for different materials and applications, their role in belt conveying systems is of particular significance. When feeding, the chute acts to accelerate the granular material and direct it onto a faster moving conveyor. In other instances, transfer chutes direct the flow from one conveyor to another and discharge chutes ensure the stream that leaves the

Simulations in Bulk Solids Handling: Applications of DEM and other Methods, First Edition.
Edited by Don McGlinchey.

Figure 2.1 Transfer chute examples.

conveyor will fill a bin, stockpile, truck, or wagon. To direct the material flow from one conveyor to another, often several transfer chutes and other equipment (e.g. rock boxes and controllable flaps) are installed in a single transfer station. In all applications, the chute will be designed according to specific performance criteria. These criteria typically require adaptive designs to handle bulk solids of varying characteristics and for a range of operating conditions. Consequently, the design must be unique for the specified application, clearly defined, and designed based on the material characteristics of the bulk solids handled. For operational reliability, it is essential the design of the chute is integrated into the early stages of plant or facility layout design, alongside other components that include conveyor and steel structure. The most common design criteria call for the following:

- Loading the outgoing conveyor at velocity that matches belt speed;
- Selection of the wall lining material to optimise flow;
- Chute inclination to prevent build-up;
- Minimising propensity for blockage, generation of dust, and wear of internal surfaces;
- Centralising the flow onto conveyor being loaded to prevent belt mistracking; and
- Sufficient skirt-board arrangement to prevent spillage.

2.2 Practicalities – Balancing Conflicting Objectives

Traditionally, the bulk solids handling industry relied on the copy and paste approach: that is – copy what works elsewhere. With the advance of science in this field, transfer chute design has evolved to find its place within conveyor and more broadly mechanical materials handling design. In this space, engineering design specifications tend to group the design criteria thereby preventing case specific requirements of each application. Effectively, the product then becomes conflicted in its purpose and often fails in its anticipated performance. One common example of such a conflict follows:

> The transfer chute shall ensure the in-line component of the material stream loading velocity matches the belt speed of the receiving conveyor.

This is conflicted by the following:

> Transfer chute inclinations and valley angles must be at least 60° to the horizontal.

Such conflicted design requirements can result in compromised chute and/or conveyor belt performance. When viewed according to practicalities, guidelines for transfer chute design [1] are a collage of good intentions and contradictions:

- Transfer head height and geometry must sometimes conform to site requirements such as minimum chute inclinations or specific shape (e.g. curved 'hood' or 'spoon');
- The most favourable wall lining seldom offers low friction and high wear resistance; wear liners can also be expensive;
- Minimising wear can contradict the conditions for minimising build-up (e.g. use of ledges instead of smooth liners);
- Reducing velocity by abrupt impact to minimise wear may increase the likelihood of conveyor belt mistracking and/or spillage (e.g. a vertical deflector compared to a curved deflector);
- Layout design that includes considering conveyors, maintenance hatches, and steel structure can lead to complex and unfavourable chute inclination and orientation;
- Steep chute angles to prevent build-up require head height, but on the other hand, large height can increase wear or dust generation due to increase in velocity, illustrated by the following:
  - steep or shallow chute and low velocity will reduce wear,
  - steep chute and low or high velocity will reduce dust and
  - steep chute and high velocity will reduce build-up.
- Matching the material stream loading velocity to the belt speed of the receiving conveyor is not practical with steep chutes due to high velocity required;
- Reducing the granular material stream velocity to reduce wear will increase the probability of build-up and blockage occurring.

In reviewing the above, it is clear that designing a transfer chute requires prioritisation and balance of design aspects to tailor it specific to each application.

2.3 The Suggested Approach

Failure to adequately incorporate the necessary balance in the previously described contradictions during early project stages will inhibit the influence that the designer has and can significantly increase the associated costs as the project evolves towards completion. This can heighten the risk to operations and the business. Obtaining this balance involves interrogating varying flow conditions, design aspects and their influence on flow velocity profiles and the resulting forces that act on chute surfaces and on the conveyors. Two main approaches to chute modelling and optimisation are widely adopted: (i) Continuum mechanics and (ii) Discrete Element Method (DEM).

The continuum mechanics approach originated with free-flowing cohesionless materials in the late 1960s/early 1970s [2] in the agricultural industry and has over time evolved to incorporate broader aspects across different resources by making appropriate assumptions [1, 3–7]. The method involves calculating flow velocities using a lumped parameter model. The model is characterised by accelerated flow and shallow bed conditions that result from frictional sliding. The input parameters include measurable physical properties such as bulk density and wall friction, but it can be expanded to include internal friction and cohesion: properties that are linked to the process parameters associated with most resource handling facilities. The continuum approach as briefly described in Section 2.4 is based on several assumptions e.g. flow conditions and stress ratios. The equations can be quickly solved. However, the consideration of complex geometries (e.g. nonconstant hood or spoon radii) results in the necessity to adopt the equations and manually decide which flow regime can be found in which section of the transfer station.

Since its conceptualisation in the field of rock mechanics [8], DEM has been used to study many different applications including sampling and blending [9, 10], silos [11], feeders [12], conveyors [13–17], and vibrating screens [18]. The DEM offers a more detailed and general approach to study the behaviour of granular materials in transfer chutes due to the consideration of individual particles and their physical interaction. Provided accurate simulation parameters are selected, the DEM analysis can also become quantitative regarding e.g. particle velocities, forces, mass flow, or wear rates. However, similar to the continuum approach, the DEM requires assumptions and often idealisation regarding the particle properties and the simulation parameters. These limit the use of the DEM. Furthermore, the DEM requires much higher calculation times than the continuum approach.

Taking into account the benefits and limitations of each approach, the practical way to transfer chute design can be summarised as:

- Develop design criteria based on general design principles specific to each application;
- Use the continuum mechanics approach to establish a baseline and quick understanding of transfer chute design functionality;
- Use DEM simulations to study the flow behaviour in more detail and refine the design.

Following is a description of the continuum and DEM methods and how they are used in transfer chute design.

2.4 Basic Design Principles – The Continuum Approach

Prior to considering flow through the chute, it is important to first understand the discharge process which starts with the burden being transported. The profile of the burden on the conveyor belt can be considered either circular [19] or parabolic [20].

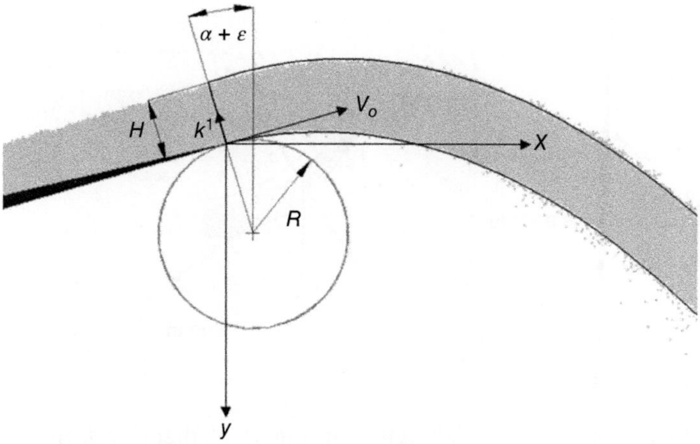

Figure 2.2 Trajectory model.

While many authors have studied the discharge trajectory going back almost 100 years (e.g. [19, 21–23]), for fast speed trajectories which typify high-throughput systems (i.e. burden discharge at the point of tangency between the conveyor belt and the drive pulley), the burden height on the belt at discharge, H, can be assumed equal to that at the final fully troughed set of idlers prior to discharge. Here, the zone between the final idler set and the drive pulley is the transition zone of the conveyor [24]. The discharge trajectory is influenced by the length of this zone, the inclination of the idler rolls, the speed of the belt, and the diameter of the drive pulley, in addition to bulk solid properties which include size distribution, density, moisture, cohesion, and friction [5]. The angle of discharge is dependent on the transition zone, and in general, it is slightly higher than the inclination of the conveyor [19]. The discharge model is illustrated in Figure 2.2.

Discharge occurs at the point of tangency if the following condition is satisfied:

$$\frac{v_0^2}{\left(R+\frac{H}{2}\right)g} \geq 1 \tag{2.1}$$

where $R + H/2$ is the centre-to-centre distance from the drive pulley to material burden, v_0 is the belt velocity, and R the pulley radius. The equation below can be used to plot the discharge trajectory:

$$y = \frac{g}{2v_0^2 \cos^2(\alpha+\varepsilon)}x^2 - \tan(\alpha+\varepsilon)x \tag{2.2}$$

If multiple streamlines are considered, then parallel parabolic curves can be assumed [6]. Following discharge, a continuum mechanics-based approach can be used to analyse velocities through the transfer chute.

The continuum approach [20, 25–27] assumes accelerated conditions, where the bulk solid is in continuous frictional sliding contact with the chute bottom and side walls during flow. This is illustrated in Figure 2.3, for both an inverted curved impact plate (hood) and a curved chute (spoon). In both instances, acceleration occurs when

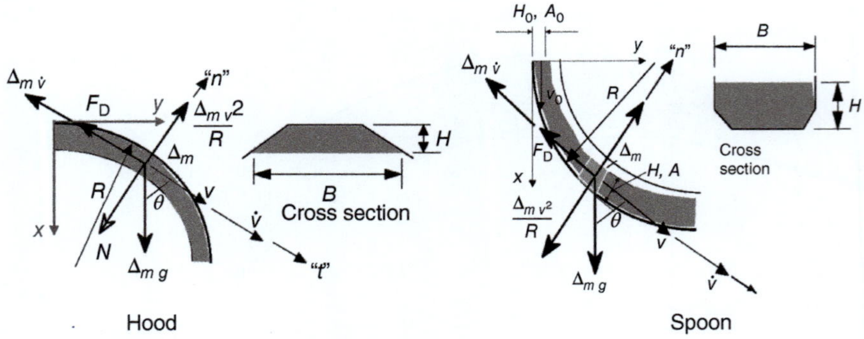

Figure 2.3 Chute flow models.

the stream velocity increases from an initial value v_0 during which there is a decrease in the stream height. When the velocity of the stream attains a maximum, the drag force F_D dominates and the flow decelerates, resulting in increased stream height and decreasing velocity. Excessive flow retardation can propagate quickly, especially in an enclosed chute, which can lead to blockage occurring, and therefore, close attention must be given to the velocity profile within the chute.

In general, chute cross-sections are primarily rectangular; however, they can also be curved. The curved chute may be rotated to prolong wear life, while the rectangular cross section can easily be lined with wear-resistant materials. The use of chamfers in the corners minimises the build-up propensity. To maintain 'fast' flow or sufficient acceleration, the chute cross-section must be designed such that the bed depth, H, is maintained smaller than the width, B, during flow, i.e. $H \leq B/2$. The height-to-width ratio H/B may differ depending on the situation, e.g. minimising dust, wear, or build-up.

In these shallow bed conditions, the highest F_D acts on the chute bottom. Assuming a lumped parameter approach where continuity is maintained, a varying equivalent friction, μ_E accounts for velocity, v, and H variation according to the coefficient of sliding friction, μ between the bulk material and the chute walls. The expression for μ_E is given by [25, 26]:

$$\mu_E = \mu\left(1 + K_V\frac{H}{v}\right) = \mu\left(1 + \frac{C}{v}\right) \tag{2.3}$$

where $\mu = \tan\varphi$; φ = chute wall friction angle; K_V = normal pressure ratio (typically 0.4–0.6) and

$$C = K_V\frac{v_0 H_o}{B} \qquad \text{(i.e.based on initial conditions)} \tag{2.4}$$

The equations of motion for each flow model may be derived for the coordinate system relevant to each type of chute. For the spoon:

$$\frac{dv}{d\theta} + \mu_E v = \frac{gR}{v}(\cos\theta - \mu_E\sin\theta) \tag{2.5}$$

For the hood, there is a slight change as the gravitational component is inverted:

$$-\frac{dv}{d\theta} + \mu_E v = \frac{gR}{v}(\cos\theta + \mu_E\sin\theta) \qquad \text{for } \frac{v^2}{Rg} \geq \sin\theta$$

If it is assumed that bulk density remains constant, H may be calculated from the cross-sectional area, A:

$$A = \frac{A_0 v_0}{v} \tag{2.6}$$

where A_0 and v_0 are the initial conditions of the stream cross-sectional area and velocity.

While the spoon flow model can be applied to 'soft' loading, chutes can also follow a straight inclined profile. Here, the material stream at a distance, s, down the incline can be calculated by:

$$v = \sqrt{{v_0}^2 + 2as} \tag{2.7}$$

where acceleration $a = g(\cos\theta - \mu_E \sin\theta)$ and θ is the inclination to the vertical.

For internal shear flows, exhibited with cascading ledges or rock-box type chutes, μ_E can be related to the effective angle of internal friction, δ, of the bulk material handled. In this instance, the equivalent friction is defined by $\mu_E = \sin\delta$. With any sloped flowing surface, it is important that the inclination angle at exit or discharge is sufficient to ensure flow under all operating conditions. Inherently, minimum chute inclination correlates to the coefficient of static friction or equivalent friction (i.e. internal angle of friction for the case of a rock-box chute).

The velocity profile calculated using the equations above can be used to approximate the abrasive wear of the chute surface. Chute wear occurs due to both impact and abrasion. Impact wear arises when there is a change in direction, and it is dependent on the impact angle, velocity, and density of the particles. Abrasive wear, on the other hand, depends on normal pressure, the coefficient of sliding friction and velocity [25, 26] and can be expressed in a non-dimensional form:

$$N_{WR} = \left(\frac{v^2}{Rg} \pm \sin\theta\right) \tag{2.8}$$

Here, the negative value is used in hood analysis. For wear on the belt, the impact pressure, P, is given by:

$$P = \rho \cdot v^2_{\text{Normal}} \tag{2.9}$$

where ρ is the density of the bulk solid material. The parameter, W_a, which expresses the rate of abrasive wear on the belt [25, 26] can be calculated according to:

$$W_a = \mu \cdot \rho \cdot v^2_{\text{Normal}} \, (v_{\text{Belt}} - v_{\text{Loading}}) \tag{2.10}$$

where μ is the coefficient of friction between the material and the belt, v_{Normal} is the normal velocity component at loading, v_{Loading} is the velocity component in the direction of belt travel, and v_{Belt} is the outgoing belt velocity. Both the non-dimensional wear profile, N_{WR} and the abrasive wear parameter, W_a can be used to differentiate between designs and enable optimisation of chute profiles with respect to abrasive wear. More details on this and on impact wear will be presented later in the chapter.

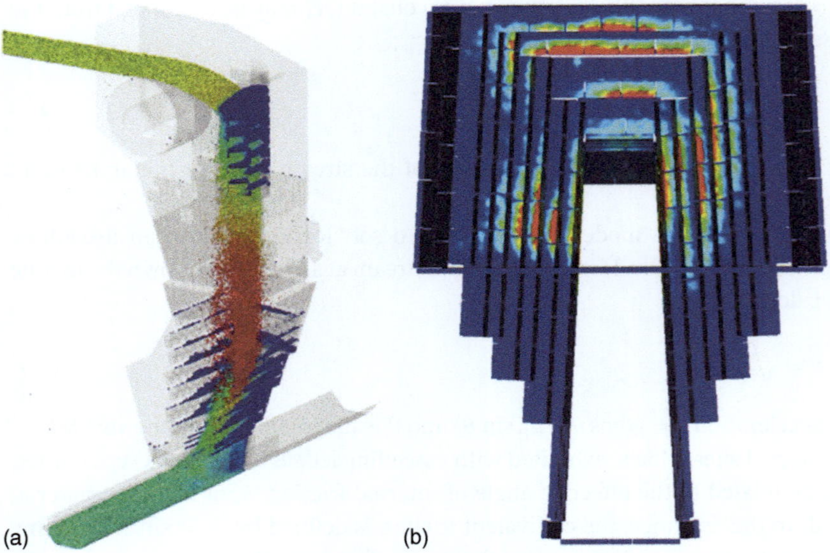
(a) (b)

Figure 2.4 (a) Flow and (b) qualitative wear pattern analysis in a DEM simulation of a transfer chute.

2.5 DEM Simulation of Transfer Chutes – General Comments

The application of DEM to solve industrial problems has increased in line with the rise in computing resources, with the most common application being transfer chutes with many publications e.g. [3, 4, 23, 28–32]. With the application of DEM simulation, the user can analyse the bulk material behaviour inside the transfer chute by:

1. Visualising the flow patterns see Figure 2.4a. It is typical for particles to be coloured by their velocity (e.g. blue for low and red for high velocity). Common features to look for in a transfer chute analysis are regions where possible flow choking or blockage can occur as well as any possible non-central loading on the receiving belt.
2. Wear patterns on surfaces are viewed – this can include both transfer chute and conveyor belt surfaces – see Figure 2.4b. It is typical for surfaces to be also coloured by their wear value (e.g. red being the maximum and blue being the minimum). Both impact wear and abrasive wear numerical data can be analysed.
3. Forces due to bulk solid interaction with surfaces can be extracted from the simulation data.
4. Breakage of particles is directly simulated. A detailed calibration process is used to quantify the breakage model parameters such that the DEM simulation can accurately predict the particle breakage (first the propensity to break and then broken particle size distribution is modelled).

One of the main advantages of DEM is the ability to simulate many design iterations without the need to fabricate and install each transfer chute variation. This is particularly useful when wanting to investigate minor design modifications, variation in operating conditions, or bulk solid characteristics. Evaluating a design change in the DEM environment is quick, and a new design can be typically evaluated within hours of calculation time.

Before a successful DEM simulation can be undertaken, it is first necessary to select the particles which represent the bulk material in the simulation. The interaction between the particles, and hence, the macroscopic behaviour of the simulated bulk material is defined by contact models with specific parameters. Most of the parameters, however, are not directly measurable. They include Young's modulus, coefficient of restitution, damping, particle/particle, particle/boundary sliding friction, Poisson's ratio, and rolling friction. Selecting accurate input parameters requires calibration, which is discussed in detail in Chapter 1.

The calibration procedure is typically undertaken using bench-scale laboratory tests to assist in selecting simulation parameters that represent physical properties. It is a widely reported field of research in its own right e.g. [33–36]. Application-specific calibration can also be undertaken on site. This can include site data (e.g. photographs, videos, belt thickness measurements, downtime information, and anecdotal evidence from maintenance crews and engineers) and laser scanning or thermal imaging of site equipment (e.g. build-up within the chute or profile of worn surfaces) including the geometry that is actually in place [37]. Field-based calibration and flow verification require filtering important information in view of aims and targets to prescribe a solution. Further, the application-specific calibration can be much more time-consuming than the classical calibration based on laboratory tests.

The optimisation task of the transfer chute design arrives in industrial practice often together with the two following boundary conditions: (i) Proper calibration requires time which in industrial application is seldom available and (ii) the bulk material parameters are often changing, and it is difficult to get a sample of the worst case material. Consequently, sufficient parameters need to be selected and a solution developed quickly. In such an instance, the 'problem' is modelled, and a sensitivity analysis undertaken so that a design solution can be proposed to qualitatively alleviate the problem to some extent. Typically, several transfer chute designs are involved in this sensitivity analysis, and the comparison of the simulation results allows the identification of an optimised design. Using wear as an example, the solution may involve producing a design that qualitatively reduces the average magnitude of wear by some factor. For build-up, the solution may involve a design that results in increased simulation time for blockage to eventuate. Through such analysis, it is rare that full-scale industrial behaviour can be quantitatively replicated; however, the use of a calibrated model can provide confidence and/or certainty in the anticipated outcomes.

In the case of DEM, practicalities include the scale-up of particle size, idealising particle shape (e.g. modelling the granular material as spheres), modelling only a short duration of flow (e.g. 60–120 seconds); all as a consequence of the time requirements to conduct the simulation. These simulation aspects are in large part influenced by a number of considerations:

- Can the mechanics associated with particle/particle and particle/boundary interactions underpinning the flow mechanisms be adequately captured (e.g. flow that occurs due to granular material sliding along the chute surface or shearing/failing internally)?
- Can the scale of the flow mechanism be modelled with adequate resolution (e.g. slow build-up of fines over time in practice simulated by large spherical particles in a relatively short simulation)?
- Can the throughput actually be modelled with realistic particle sizes? In high-throughput systems (10–20 000 tph), it is impractical and sometimes not possible to include very small particles. Conversely, at low throughput, consideration also needs to be given to the scale between particle size and the cross section of the granular material stream.
- What simulated condition or sets of conditions are necessary to replicate the specific physical behaviour that is sought? Can the condition be replicated in the simulation environment? What assumptions need to be made for the results to be meaningful?
- What is the acceptable simulation time? Is assessment of flow at steady state on the outgoing conveyor sufficient or is the flow mechanism under investigation time-dependent?

While calibration is widely reported, academic literature documenting the application and practical use of DEM in the context of transfer chute design is gaining attention in parallel with industry uptake (e.g. [1, 3, 4, 6, 7, 30, 37–44]). In the following, we will give an overview of the most important findings regarding the use of DEM to optimise transfer chute design regarding the criteria of general flow, generated belt mistracking, build-up and blockage, wear, and dust generation. This should familiarise the reader not only with the possibilities of the DEM analysis but also with its limitations. We will show to what extent the DEM is able to predict:

- the material flow inside the transfer chute,
- the forces occurring during the transfer,
- the eccentric belt loading as a cause for belt mistracking,
- the material build-up in chute which can cause blockage,
- the wear inside chutes,
- dust emissions caused by the material flow.

The chapter is summarised by a section about balancing the conflict which can arise due to the different design aims.

2.6 General Flow Analysis and Validation

The following sections describe interrogation and validation of the continuum mechanics theory and the DEM approach to transfer chute design and analysis.

2.6.1 Discharge and Trajectory

Analysis of the bulk solid behaviour due to the transition geometry of the conveyor belt cross section in the zone from the last idler set to discharge is generally constrained to two dimensions. In practice, the transition geometry (illustrated in Figure 2.5) influences both the dispersion and discharge trajectory in, and perpendicular to, the conveyor belt travel direction. Therefore, discharge trajectory requires analysis in three dimensions [5].

A short transition results in more lateral dispersion, and free-flowing materials will disperse more compared to cohesive materials. Depending on moisture content, the burden profile for cohesive materials also exhibits a tendency to break up according to the cross-sectional transition geometry and material properties, namely, the internal angle of friction. This is illustrated in Figure 2.6. A comparison of free-flowing and cohesive cross sections, experimental (top), and simulated using DEM (bottom), is shown in Figure 2.7.

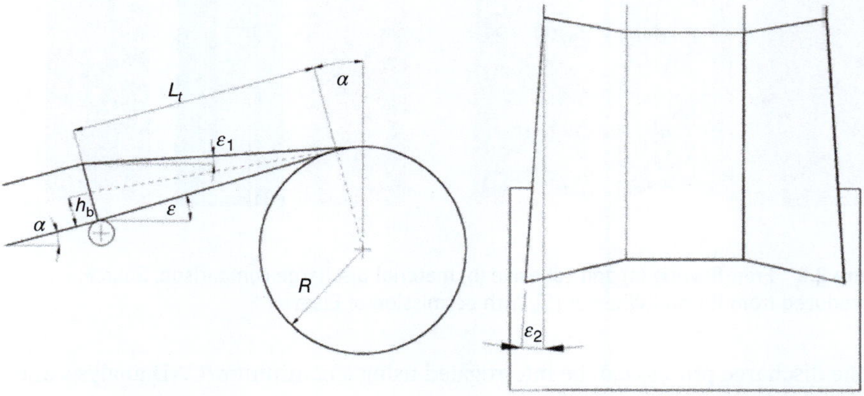

Figure 2.5 Conveyor belt transition geometry. Source: Reproduced from Ilic and Wheeler [5]/With permission of Elsevier.

(a) (b)

Figure 2.6 Discharge of free-flowing and cohesive materials (a to b). Source: Ilic and Wheeler [5]. Adapted from Elsevier.

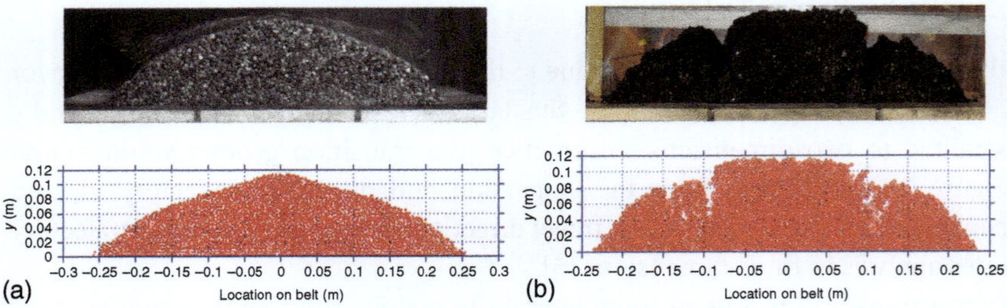

Figure 2.7 Free-flowing (a) and cohesive (b) burden at discharge - comparison of experimental and simulated results. Source: Ilic and Wheeler [5]. Adapted from Elsevier.

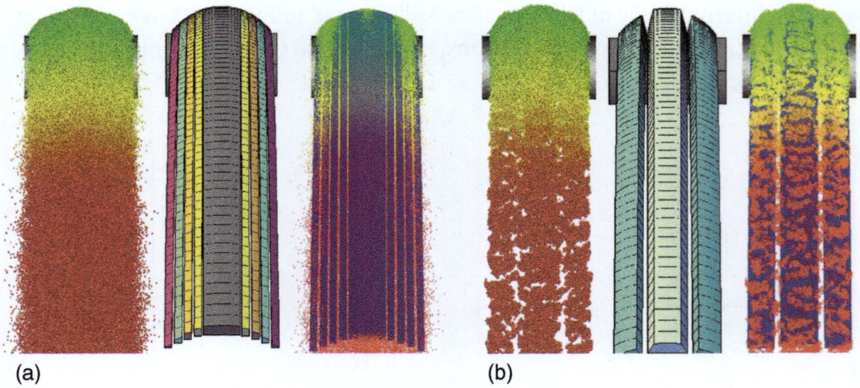

Figure 2.8 Free-flowing (a) and cohesive (b) material discharge comparison. Source: Reproduced from Ilic and Wheeler [5]/With permission of Elsevier.

The discharge process can be interrogated using a continuum/CAD analysis and conveyor belt velocity [5]. The approach approximates average bulk behaviour for the sections modelled, while DEM can actually simulate the resulting interactions at a much finer level of detail (i.e. closer to actual particle size handled). Analysis and comparison between the continuum and DEM modelling approaches are shown in Figure 2.8.

2.6.2 Accelerated Flow

In general, for fast-flowing conditions (accelerated flow), spoon flow and loading analysis shows good agreement between the continuum and DEM approaches [6]. This is illustrated in Figure 2.9. The comparison highlights that particle parameters simulated with DEM are of less influence on the overall flow through the transfer at faster average material stream velocities.

While the continuum mechanics approach can provide a quick overview of chute functionality, it is limited by its underlying assumption of fast-flow conditions, and due to this, it cannot accurately describe transitional flow regimes (e.g. fast to slow flow transition characterised by rapid deceleration and/or high momentum

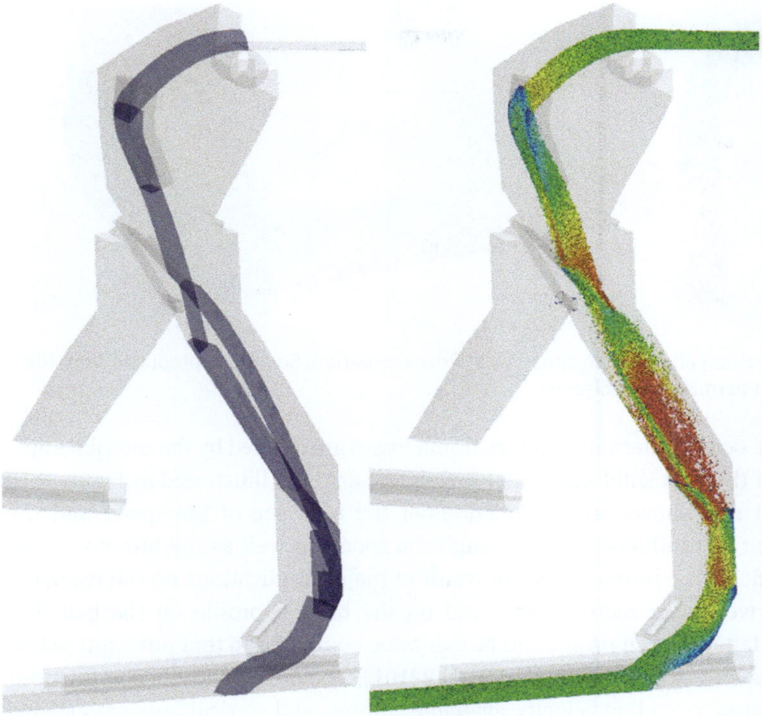

Figure 2.9 Comparison of continuum mechanics and DEM flow simulation. Source: Reproduced from Ilic et al. [6]/With permission of Elsevier.

losses due to abrupt impact). In this instance, DEM analysis enables much greater resolution and insight.

2.6.3 Internal Shearing Flow

When material flow is constrained and forced through a smaller space than optimal, such as in stacker reclaim central chutes (e.g. [4, 7]), different or combined modes of flow are experienced. Once again, an excellent agreement between the continuum mechanics and DEM approaches is obtained in fast flow conditions (such as flow through the hood – illustrated in the left image in Figure 2.10). However, significant discrepancy can arise for areas of high deceleration, such as the flow through the spoon (also illustrated in the right image in Figure 2.10). In such an instance, flow visualisation using DEM can assist in altering the underlying assumptions, based on which, the continuum approach can be modified to suit. Subsequently, this enables a different mode of flow to be analysed using both approaches, such as that governed by internal shear due to the formation of a quasi-surface formed by slow moving particles (illustrated by the dark blue region in Figure 2.10).

Such revised assumptions can enhance understanding of the mode of flow and the failure surface. In this case, rather than a velocity profile defined by frictional losses due to failure at the interface between the material stream and the chute

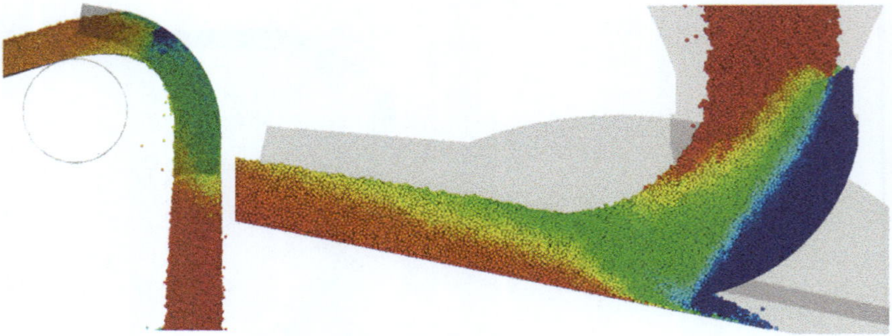

Figure 2.10 Hood and loading spoon DEM flow simulation. Source: Reproduced from Ilic et al. [7]/With permission of Elsevier.

walls, failure occurs internally, and frictional losses are defined by the internal angle of friction of the bulk solid material. Further, the analysis illustrated in Figure 2.10 on the right also shows how interactions at the interface of the spoon and the receiving belt can influence flow through the spoon as well as the burden on the receiving belt. These interactions can result in major implications on the receiving conveyor drive power requirements and on the burden profile on the belt. For example, a flatter burden profile can be exhibited compared to that pursuant to belt conveyor design guidelines (e.g. [19], DIN 22101). The burden profile is influenced by material stream loading velocity, available volume, and vertical forces associated with the falling material stream.

2.7 Quantitative Validation of Forces in DEM Simulation

As DEM enables the analysis of contact forces during the entire simulation and for all surface boundaries, it is possible to use this information to optimise designs. The acting forces on the transfer chute surfaces caused by the material stream play a major role in its structural design. High forces require a stiffer design and reinforcement of the steel structure which results in a higher mass of the whole transfer chute. This causes higher requirements for the surrounding steel structure and results in a higher cost. The analysis of forces is also the basis for the wear analysis (see Section 2.10) which allows a lifetime prediction of the wall liners and the conveyor belt.

For the validation of forces occurring in transfer stations, Katterfeld et al. [30] compared DEM simulation results with the experimental data from a heavy-duty transfer station in one of Germany's large lignite mines (see Figure 2.11). The bulk material tested was very coarse clay iron ore. As expected, the measurements mainly depended on the mass flow rate. The force measurements on the impact wall in DEM and experiment showed generally a good correspondence. The measurement of the impact forces on the idler sets showed overall forces with the same magnitude. However, due to the idealisations of the simulation model regarding e.g. the rigidity of walls which were used to model the conveyor belt, several differences were found between the experimental and simulated results.

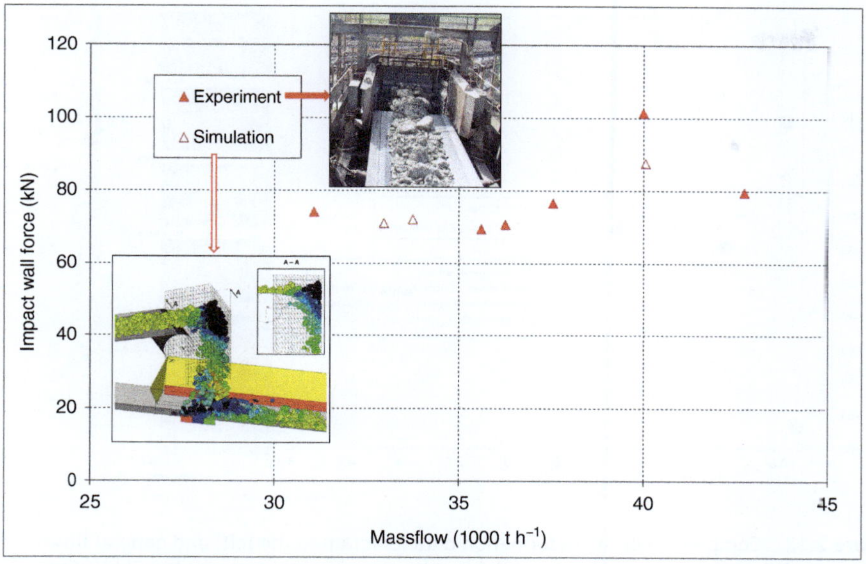

Figure 2.11 Comparison of force measurements on a vertical impact wall in a large-scale industrial transfer chute and DEM simulation. Source: Adapted from Katterfeld et al. [30].

These early results showed the problem, that for more detailed force prediction, the interaction between machine parts and the bulk material needs to be considered in the simulation model. Such a detailed analysis is necessary if the dynamic load and behaviour of parts is required. Due to the fact that in pure DEM simulation, the geometric boundary conditions – meaning the walls of transfer chutes in our application – cannot deform or move depending on the loads resulting from the flow of bulk material, a coupling of the DEM with multibody and/or Finite Element Method (FEM) simulations is necessary.

Dratt and Katterfeld showed in [45] how precise a coupled DEM–FEM simulation can predict deformations (and hence forces) of a steel sheet which acts as an obstacle in a free-falling bulk material stream (see Figure 2.12).

Coupled DEM–FEM simulations are especially important for the analysis of belt impact forces. Such simulation can support the proper design and selection of impact idlers or impact bars. If parts of the transfer station should move depending on the load (e.g. an impact wall or a flap which moves into the material stream and redirects or diverts the flow, impact chain curtains, or idler garlands), a coupling with multibody simulation is necessary as described in more detail by Richter et al. [46].

2.8 Belt Mistracking

One problem often related to the material flow in the transfer chute is belt mistracking or runoff. Belt mistracking as shown in Figure 2.13 is the unwanted sideways movement of the belt [38]. This is one of the most common operational problems

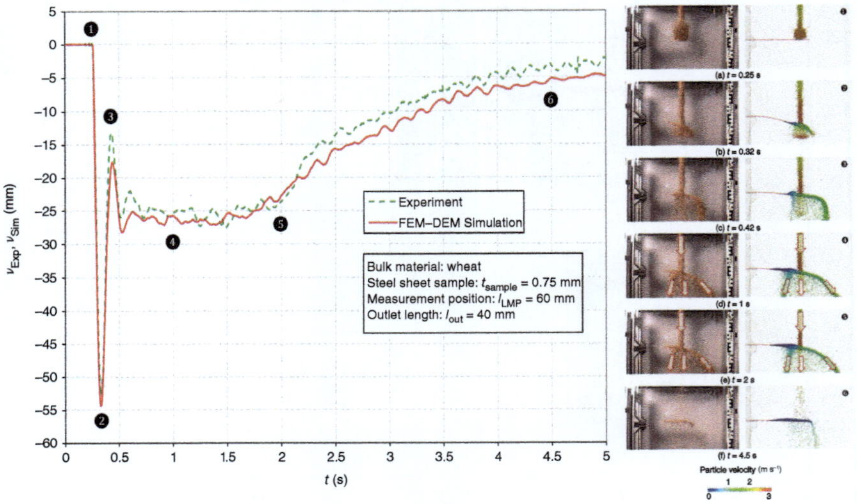

Figure 2.12 Comparison of the deformation data (diagram to the left) and general flow observation (image sequence to the right) from the experimental test and the FEM-DEM simulation with wheat (6 kg), a steel sheet sample thickness of 0.75 mm, an outlet opening length of 40 mm, and a measurement position of 60 mm from the right edge of the steel sheet. Source: Dratt and Katterfeld [45]/with permission of Springer Nature.

Figure 2.13 Off-centred loading of bulk material on a belt conveyor causes mistracking of the belt. The belt mistracking can be recognised via the comparison of the belt coverage of the side rollers and the location of the material on the belt.

of belt conveyor plants, ports, mines, and materials handling facilities, causing a significant amount of supply and production failures and operational downtime.

According to [47, 48] the most common reasons for belt mistracking are:

- Non-central belt loading,
- material build-up on idlers or pulleys,
- insufficient belt tension,

- extreme weather such as side wind or heavy rainfall,
- misalignment of idlers or head and tail pulley,
- belt cambers due to insufficient manufacturing accuracy.

In [49], it is described that the off-centred loading of the bulk material onto the belt is one of the major causes of belt mistracking. Here, material is loaded eccentrically onto a centred belt due to incorrect settings or a poor design of the transfer chute. The weight force of the eccentric bulk material results in a belt deflection which causes a sideways movement. A self-intensifying eccentric belt movement results if the belt is not centred before the transfer chute, as the bulk material will be loaded eccentrically onto an already off-centred running belt. This can be solved by redesigning the transfer chute that loads the belt. To investigate the magnitude of the bulk material pushing the belt sideways, DEM simulations can be used. Otto published in [38] a DEM simulation to show the influence of the off-centred loading of bulk material on the tangential force on a (fixed) belt. The simulated bulk material is transferred from a feeding belt to a receiving belt via a vertical impact plate. The transfer angle is constant 90°. The position of the impact wall changes in the simulation. The initial position during 0–4 seconds simulation time causes a centred loading of the receiving belt as shown in Figure 2.14 left. During the movement of the impact wall in y direction, the centre of gravity of the bulk material is shifted in positive y direction as well. The resulting forces on the different receiving belt faces (left, centre, and right) can be measured in the simulation over time. It was found that the tangential forces (acting in y direction in Figure 2.14) depend on the eccentric load of the bulk material. Most significantly, the tangential forces on the centre face of the belt determine the direction of belt mistracking. In the given example with an eccentric belt load to the right (in $+y$ direction), the resulting tangential force acts in the opposite direction to the left. This was exactly observed in reality e.g. in Figure 2.13.

By using the described analysis of forces acting on the belt, changes in the design of the transfer chute can be directly examined. Hence, an optimisation of the transfer chute design can be realised to minimise belt mistracking.

2.9 Simulation of Build-Up and Blockage of Cohesive Material

One key task regarding the design of transfer chutes is the prevention of blockage. With a continuous volumetric flow rate according to Eq. (2.6), the cross section of the transfer chute must be chosen based on the speed of the bulk material stream or vice versa. The size of the selected cross section must also take into account the maximum size and shape of the particles to prevent mechanical interlocking. As described in Section 2.6 the DEM is the ideal tool to predict the speed of the material flow and hence identify deceleration zones which can result in an increased risk of blockage or overfilling as shown in Figure 2.10.

However, a concern even for DEM simulation is the prediction of long-term build-up. In reality, a transfer chute does not always block immediately. It is a

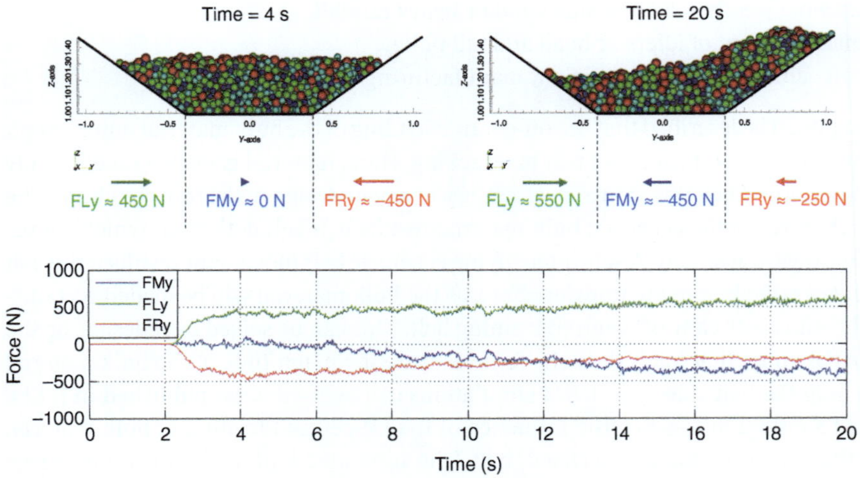

Figure 2.14 The change of centred and off-centred loading of bulk material on a simulated belt conveyor causes mistracking of the belt due to the change of tangential forces acting on the belt (particles are coloured by their radius). The diagram below shows the timely change of tangential forces in the three segment of the belt inclined. Source: Otto [38]/CC BY 4.0.

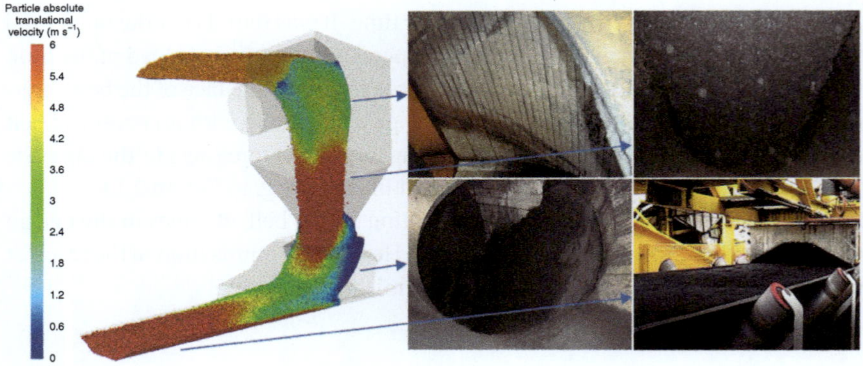

Figure 2.15 Comparison of DEM simulation results of a hood and spoon transfer chute with photos of the real transfer chute on site. The zones of low particle velocities correspond spatially very well to zones of light build-up at the hood and accumulated material at the spoon. Source: Refs. [4, 7, 30, 50]/Adapted from Elsevier.

common problem that material sticks and accumulates at chute walls where a low particle velocity can be found. Typically, this occurs in the corners where the movement of particles is inhibited from two sides and in situations where chute walls have a low inclination as shown in Figure 2.15: an example of DEM simulation of a chute handling coal at high throughput. In this case, photos from a site visit during operational downtime are available, so that a direct comparison between slow regions in the predicted material stream and accumulation zones on the hood and spoon can be made.

This build-up effect is increased if high adhesive forces occur between the material stream and the lining material of the chute wall. To illustrate, Figure 2.16 shows

Figure 2.16 Material flow of wet gypsum (moisture content 23%, bulk density 950 kg m^{-3}, particle size <0.2 mm) from a belt conveyor and along an inclined chute out or Perspex®. Steady-state operation (a), and remaining material after stopping the belt conveyor (b).

conveying tests with wet gypsum and the remaining bulk material at an inclined chute after stopping the material transport. Carr et al. showed [51, 52] that the sticking of material can even occur on vertical chutes. During a pause in operation, the remaining material on the chute may dry or harden or – if larger quantities remain on the chute – it may even show time consolidation effects. If the transport starts again, the hardened build-up will disturb the flow of 'fresh' bulk material. Hence, the remaining material changes the geometric boundaries for the material flow. Due to the lower inclination angles around the build-up, this change negatively influences the material flow. Therefore, it is easy to imagine that after a series of transport and pause phases the build-up on the chute will grow.

The described accumulation or build-up can cause a decrease of the available cross section for the material flow, which then causes a sudden chute blockage. To reach such a critical build-up or critical cross section it usually takes hours, days of operation or even longer. The process is dangerous if the critical zones are not known as such and hence are not maintained regularly or even worse, cannot be maintained due to difficult access from the outside.

The build-up risk depends on the properties of the bulk material. Cohesive and adhesive material – so-called wet and sticky materials – show a much higher risk for build-up and blockage compared to non-cohesive material. If the material is easy to compact e.g. wet powders like gypsum or limestone, the risk is even higher. The characterisation of the cohesive and adhesive properties of wet and sticky materials with focus on the influence on transfer chute design is described in detail by Carr [53].

The prediction of long-term build-up of materials in chutes in DEM simulation is a difficult task, since usual DEM simulations show the material flow only for seconds or minutes. The user needs to interpret the simulation results in this regard and must also take into account the remaining material when feeding of the material is stopped. Of course, zones with low velocity can be identified after reaching a steady state condition. However, the decrease of the flow velocity and even more importantly, the remaining material, cannot be seen if inadequate contact models are used and/or the parameters were poorly calibrated. The latter is caused due to the fact that the simulation of cohesive and adhesive materials is much more difficult than the simulation of non-cohesive materials, which is inherently a consequence of a higher effort in calibration. While the calibration of cohesionless to a light cohesive material can be undertaken according to widely accepted guidelines and suggestions from the DEM community [54], the calibration of cohesive and adhesive material is still a field of research and not fully understood, especially at the macro scale. The manifold of the available contact models for the consideration of attractive cohesive and adhesive forces does not make the task easier. Hence, the proper selection of working contact models and their parameters requires significant experience of the DEM user and/or a sensitivity study according to the general statement in Section 2.5. Therefore, the effort for such simulations is generally higher than that for standard transfer chute simulations with non-cohesive or lightly cohesive material.

The current capabilities of DEM to model build-up can be demonstrated with the example of the experimental test shown in Figure 2.16. The same test was simulated considering a liquid bridge model for the cohesive forces between the particles and a simplified Johnson–Kendell–Roberts (JKR) model for the adhesive forces between particles and walls. After parameter calibration as described by Carr [53], a qualitatively good correspondence between simulation and experiment could be reached as shown in Figure 2.17.

In [53] (using the same contact models and calibration strategy as for the tests with wet gypsum), a similar good correspondence for the material build-up was reached for wet and sticky iron ore on chute walls. However, in all cases, the remaining mass on the chute in the DEM simulation was lower than the experimentally measured build-up. This indicates that the DEM results with the used modelling technique should be interpreted as too optimistic. The reasons for this interpretation originate in the complex micro- and macro-scale behaviour of wet and sticky materials. It is known from other applications (e.g. pneumatic conveying) that moisture migration

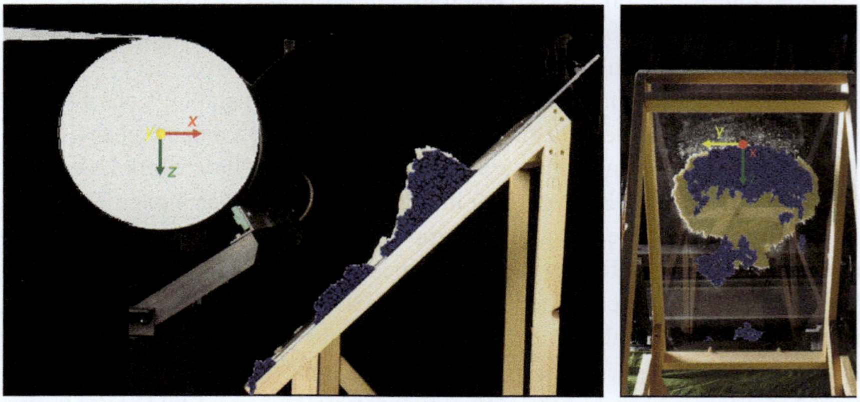

Figure 2.17 Comparison of the remaining bulk material in experiment and DEM simulation by an overlay of experimental observations and simulated results in two perspectives.

can occur in the accumulated material during initial and subsequent impacts. This may change the material properties of the build-up as well as result in additional forces at the wall. Hence, an engineer who designs a transfer chute based on a DEM simulation might not be aware of the full amount of build-up and therefore might underestimate the risk of blockage. With an increased understanding of the fundamental processes during the build-up and the further development of contact models as well as calibration strategies, it can be expected that the DEM simulation of build-up and the prediction of the blockage risk will improve in future.

2.10 Prediction of Wear

Besides the avoidance of significant spillage or the risk of blockage, the consideration of wear is a key aspect for transfer chute design [55]. Wear causes loss of volume of surfaces which are in contact with the bulk material (see Figure 2.18). Hence, wear reduces the lifetime of chute parts and of the belt if a belt conveyor is used for receiving the material stream. Due to the high mass flow rate of some belt conveyors, the velocity of the bulk material can also be high. Under such conditions, wear can badly damage and/or destroy chutes within hours. Wear-resistant liners are used to reduce the chute wear which extends the lifetime but increases costs. Increased costs are caused not only by the additional liners but also by the increase in weight in general and the more robust steel or wear-resistant engineered material construction. To increase the lifetime of conveyor belts, it is most effective to use a proper chute design because the vast majority of all conveyor belt damage is most influenced by the design of the transfer chute. Due to the fact that the belt is typically the most expensive part in a belt conveyor plant or materials handling facility, the wear aspect of transfer chute design is even more important.

Several wear types are known from which the abrasive sliding and abrasive impact wear as shown in Figure 2.19 are most dominant for transfer chutes [56]. Abrasion-induced sliding and impact wear depends on the particle movement along the surfaces. The relative sliding (=tangential) velocity and the impact angle are the most important kinematic particle parameters together with the material properties of particles and walls. Velocity and some basic particle parameters such as particle mass and wall friction coefficient can be used to calculate the energies during sliding and impact.

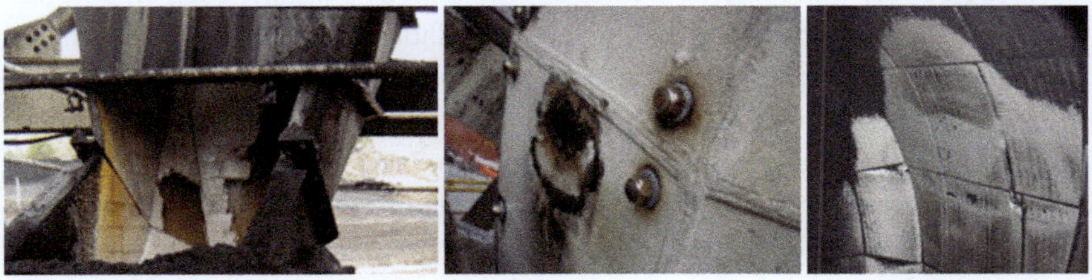

Figure 2.18 Wear of transfer chutes.

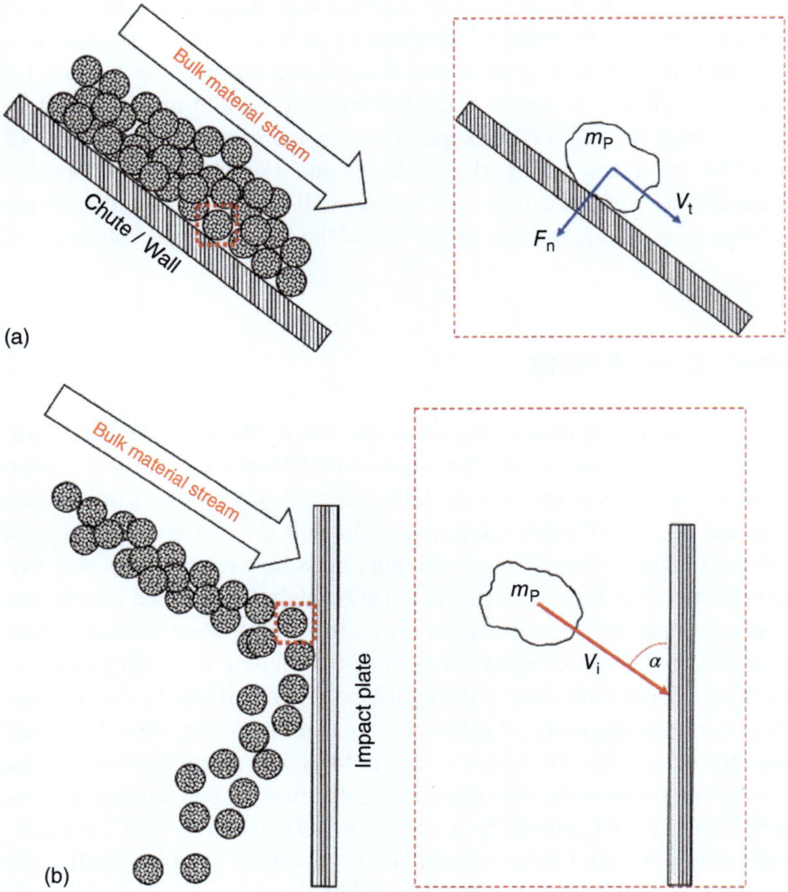

Figure 2.19 Abrasive sliding (a) and impact wear (b).

The DEM is a perfect simulation tool to analyse particle velocities and impact angles. The contact energies are often related to the time they occur or are cumulatively calculated. Hence, DEM simulations can also be used to predict wear.

Two kinds of wear analysis can be distinguished:

1. Qualitative wear analysis shows zones with high impact or tangential energies. A prediction of the wall material wear life is not directly possible. Such an analysis can be investigated by comparing several simulations with different designs or boundary conditions. An example of such a comparative analysis using Rocky DEM (ESSS) is shown in Figure 2.20.
2. Quantitative wear analysis incorporating more or less sophisticated wear models to calculate the direct volume loss and hence the lifetime of the liner materials. Different wear models are available for different flow regimes and materials. The majority of the models incorporate wear factors which must be experimentally determined considering the real flow regime, the real bulk material, and the real

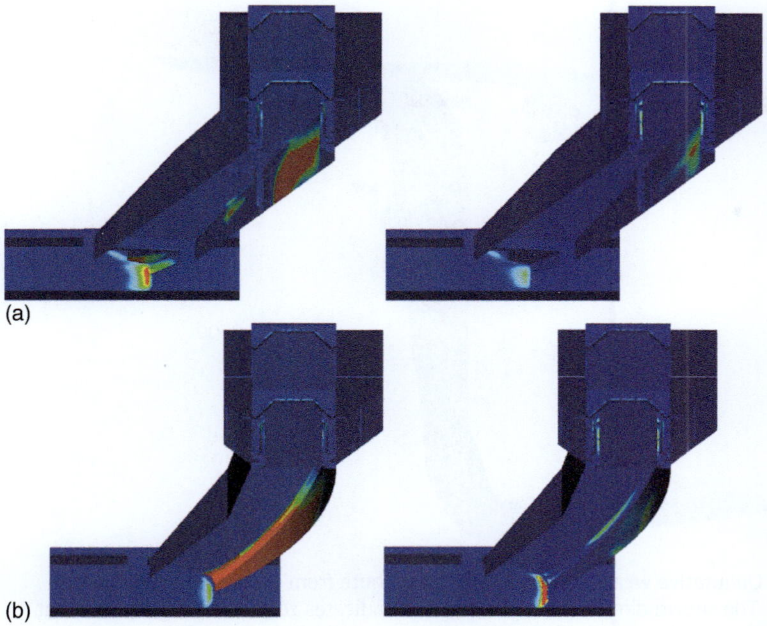

(a)

(b)

Figure 2.20 Qualitative wear comparison (from left to right) of shear and impact energy contours in [J s^{-1}] for two different chute designs, with ledges (a) and smooth curvature (b). Blue indicates zero wear and red indicates maximum wear. The other simulation parameters (mass flow rate, particle, and chute wall properties) were kept constant. Source: Ilic [1]/With permission of Elsevier.

wall liner material. If such parameters are not available, a quantitative prediction of the lifetime is not directly possible, and the results can only be used for a qualitative wear prediction as shown in Figure 2.21.

For many tasks, a qualitative wear analysis is sufficient. However, this kind of analysis does not allow the definition of exchange times for the used wear liner. For a quantitative wear prediction, the most commonly used wear models are Archard's law for sliding wear and Finnie's law for impact wear of ductile materials. For both laws, several implementations can be found in literature and the DEM software. A comprehensive overview is given by Roessler [58]. Since both wear regimes can occur at the same time at different locations of the transfer chute, a combined wear model as suggested in [58] seems to be most appropriate. For each wear, submodel wear factors have to be determined experimentally. Similar to the particle and DEM properties, the results of such experiments cannot be used directly in the DEM simulation due to the necessary idealisations of the DEM model. Therefore, it is necessary to determine the DEM wear factor by calibration. This means that the experimentally measured wear in the form of the material loss must be attained by the simulation.

An often-used experiment to determine the wear factor for abrasive sliding wear is the pin-on-disk wear test as presented by Chen in [59]. Hereby, a single DEM particle is gliding along a rotating circular or annulus surface. The bulk behaviour

Figure 2.21 Qualitative wear analysis of a transfer chute from [57] using an adopted Finnie model. The shown dimensionless wear factor indicates zones with higher wear but cannot be used to predict the lifetime of the chute. Source: Adapted from Kloss et al. [57].

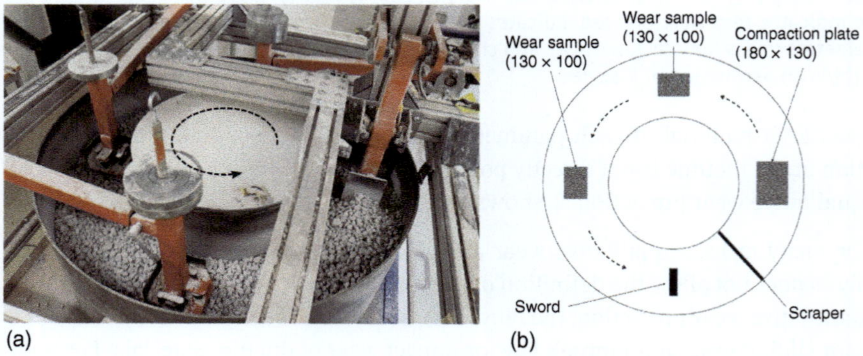

Figure 2.22 Ring wear test rig for the determination of abrasive glider wear. General view (a) and schematic (b). Source: Roessler et al. [58]/Public Domain.

of material cannot be considered with this test. Hence, [36] propagates the use of ring wear tester as described by Wiche in [60] and shown in Figure 2.22. Here, the wear is measured on two wear plates which are pressed on the rotating bulk material bed with a constant normal force. The normal force is applied via a counterbalanced cantilever beam with attached weights. The auxiliary devices sword, scraper, and compaction plate ensure a uniform mixing and abrasion of the bulk material as well as a constant material bed shape throughout the tests.

Since the ring wear tester does not allow the calibration of impact wear, an additional impact wear test as described by [61] and/or [58] and shown in Figure 2.23 can be used to determine the wear factors for the different wear models independently. In this tester, the bulk material is lifted with a bucket wheel and flows in a funnel.

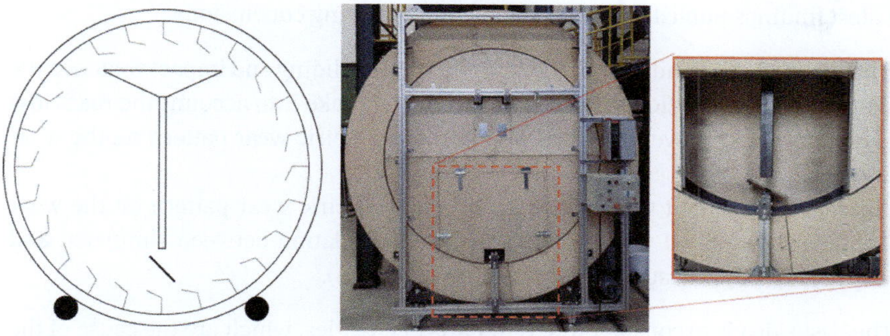

Figure 2.23 Impact wear test rig as described by [58]. A bucket wheel transports the material to a funnel through which it falls vertically down onto an inclined wall liner sample. Source: Roessler et al. [58]/Public Domain.

From here, it falls down on an inclined wall sample. The inclination of the wall sample can be changed. Rotating speed of the bucket wheel and the filling level of the buckets are chosen to generate a constant material stream which impacts the wall sample.

The different wear test rigs can be easily modelled in DEM simulations. Figure 2.24 shows the simulations of the different wear test rigs.

While wear in reality occurs over a long time (typically weeks to months), DEM simulations can only cover a time frame of seconds or minutes. This generates the question of how DEM simulation results can be extrapolated beyond a fairly small timeframe Roessler et al. show in their publications [56, 58, 62] that if small geometric changes of wear liners are assumed, a linear mass loss for ductile materials can be measured over time. Hence, the wear rate can be assumed constant. Therefore, it is possible to determine the wear rate in DEM and extrapolate this to calculate the wear of a much longer timeframe.

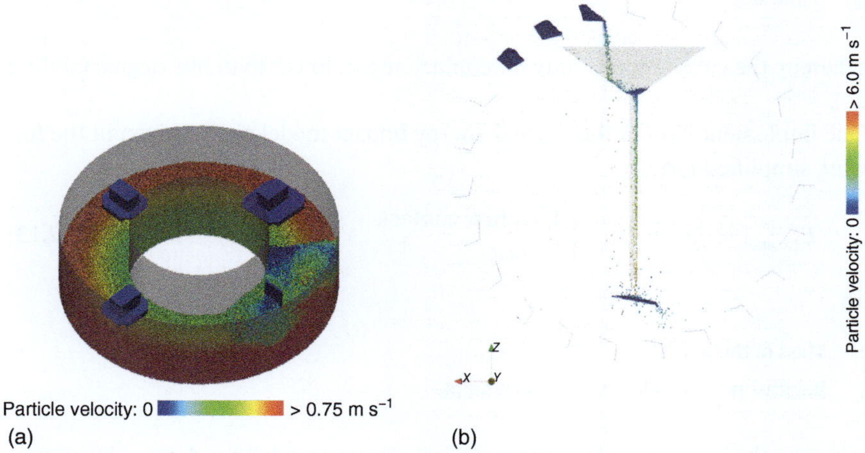

Figure 2.24 DEM simulation of the ring wear tester for sliding wear (a) and the impact wear tester with the bucket wheel (b).

Latest findings published in [58] enable the following conclusions:

1. In bulk material handling, often a combination of sliding and impact wear occurs.
2. For realistic simulation results, it is necessary to take into account the mass loss in the form of the wear rates as well as the geometric wear pattern on the wear sample.
3. Taking into account the mass loss and the geometric wear pattern on the wear sample, a combined wear model gives best correlation between simulated and experimentally measured wear (i.e. total mass loss).

The basic idea is to consider different parts of energies, which are the cause of the various forms of wear (impact and sliding) as previously reported in [1] and similar to that shown earlier in Figure 2.20. To describe the impact wear, Roessler recommends the combination of a Finnie model to describe the cutting energy part E_{Finnie} and a normal energy model to describe the deformation energy part E_{Normal} of the impacting bulk material. The pure sliding wear can be represented by the classic Archard model E_{Archard}. By weighting the different energy components with wear coefficients k_i, the resulting wear is given by

$$w = k_{\text{Finnie}}\, E_{\text{Finnie}} + k_{\text{Normal}}\, E_{\text{Normal}} + k_{\text{Archard}}\, E_{\text{Archard}} \tag{2.11}$$

with k_{Finnie}, k_{Normal}, and k_{Archard} representing the calibrated wear coefficient of cutting impact wear, deformation impact wear, and sliding wear, respectively.

The implementation of the Archard model can be given in the following simplified form,

$$E_{\text{Archard}} = F_{\text{n}} v_{\text{t}} \Delta t \quad \forall \quad \alpha \leq 0.5^{\circ} \tag{2.12}$$

with

F_{n} Normal force at contact point
v_{t} Relative tangential velocity at contact point
Δt Time step

whereby the calculation is only for contact angles lower than one degree (sliding case).

The implementation of the normal energy impact model can be given in the following simplified form,

$$E_{\text{Normal}} = \frac{1}{2} m_P {v_n}^2 \quad \begin{cases} 1 & \text{if first contact} \\ 0 & \text{else} \end{cases} \tag{2.13}$$

with

m_{P} Mass of the particle
v_{n} Relative normal velocity at contact point

whereby the calculation is only for the impact phase, which is defined by a negative velocity sign.

For implementation of the Finnie Model, different approaches in the literature exist, such as direct Finnie Model implementation considering the particle mass instead of reaction forces or the Shear Impact Erosion Model [63] in which the cutting wear is described on the basis of the shear energy. Since the different implementations have advantages and disadvantages considering insensitivity to DEM contact parameters and complexity of implementation, in [58] a direct implementation for the Finnie model is proposed,

$$E_{\text{Finnie}} = \frac{1}{2} m_P \, v_P{}^n f(\alpha_{t_0}) \begin{cases} 1 & \text{if first contact} \\ 0 & \text{else} \end{cases} \quad \forall \quad \alpha > 0.5^\circ \tag{2.14}$$

with

$$f(\alpha) = \begin{cases} \frac{2}{K}\left[\sin 2\alpha_{t_0} - \frac{7}{K}\sin^2 \alpha_{t_0}\right] & \forall \tan\alpha \leq \frac{K}{7} \\ \frac{2}{7}\cos^2\alpha_{t_0} & \forall \tan\alpha > \frac{K}{7} \end{cases} \tag{2.15}$$

whereby

v_P	Relative particle velocity at contact point
α_{t_0}	Initial impact angle at the beginning of the contact at the time t_0
n	Velocity exponent ranging from 1 … 1.5
K	Ratio of normal to tangential force approximated reciprocal of friction coefficient

2.11 Prediction of Dust

The generation of dust emissions in transfer chutes is a problem of high industrial relevance. Increasing environmental regulations force operators to prevent such dust emissions. Generally, three sources of dust emissions are known:

- Wind erosion where air movements with higher velocity generate a lift-off and a further transport of fine dust particles from a bulk material surface. This source plays a major role for the dust emissions from stockpiles, belt conveyors, or trucks.
- Impact of particles on a material bed causing splashing of material and particle attrition or even crushing which generates more fine particles.
- Dust release due to porosity changing of a bulk material stream, which is often the main reason for dust emissions in transfer chutes. Here, the air inside the bulk material is escaping together with the fine dust particles if the porosity of a material stream changes abruptly e.g. during a 90° impact. Figure 2.25 shows this effect.

Dust emissions depend not only on the outer boundary conditions (air speed, falling height, and mass flow rate) [64] but also on the dustiness of the material [65]. Beside the active dust suppression e.g. using exhausts, increasing the moisture content of the bulk material or the use of ultra-fine water fog sprayers also the passive dust prevention is possible. In comparison to active dust suppression, the passive approach does not require additional resources and can reduce dust emissions of a

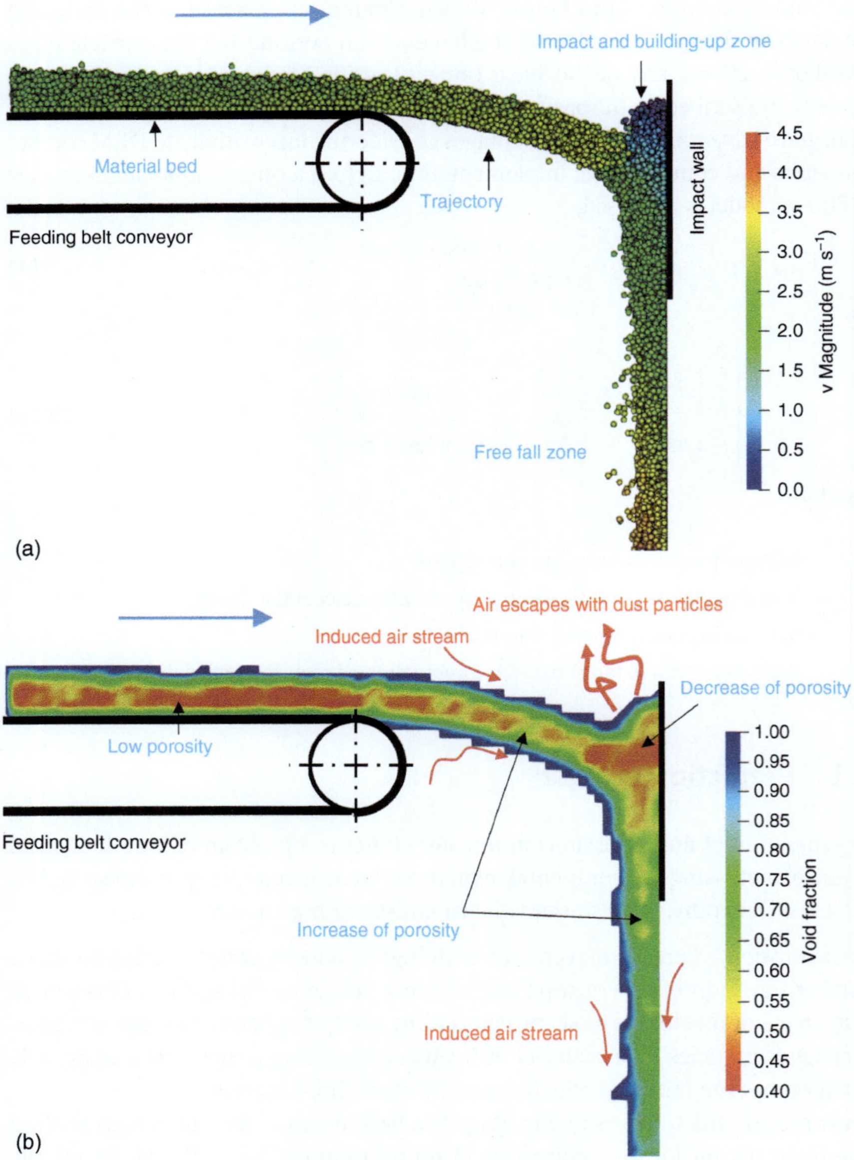

Figure 2.25 DEM simulation of a 90° transfer chute (a) and visualised porosity of the material stream (b).

transfer chute by implementing specific design aspects. The transfer chute design for low dust emissions requires the understanding of the bulk material flow and the prevention of abrupt porosity changes and impacts. Hence, three general design rules can be defined:

1. transfer chute must ensure a smooth redirection of the material flow, with no abrupt flow change,

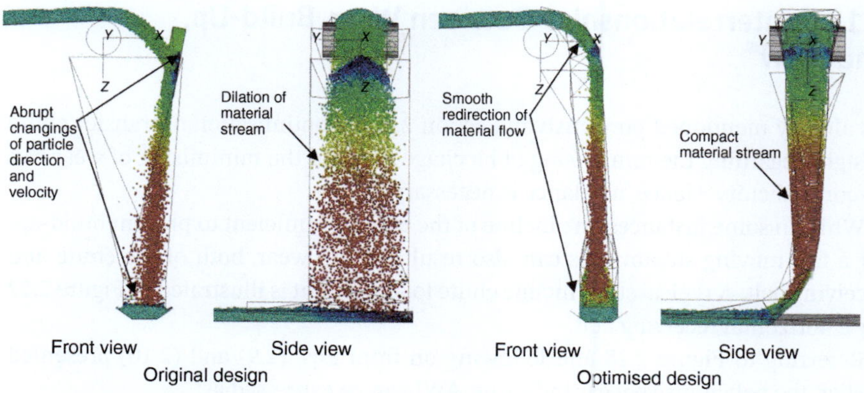

Figure 2.26 DEM simulation of the two transfer chute designs with commented dust emission aspects. Particle colour: Particle velocity: 0 to $>7\ m\ s^{-1}$.

2. prevent the accumulation of bulk material inside the parts of the transfer station (e.g. in a rock box) to reduce the impact of 'material on material',
3. keep the material flow compact and do not allow a widening-up of the material stream due to the increase of the entrained air flow.

The DEM is the right tool to predict the material flow inside transfer chutes while taking into account the properties of the bulk material and the complex geometric boundary conditions of the chute. Figure 2.26 shows how DEM simulation can be used to analyse a given transfer chute design and develop an optimised design to reduce dust emissions.

Passive dust suppression is a good solution if the bulk material properties are fairly constant and/or not changing. If this is not the case, the transfer chute design has to consider the worst-case properties to maintain the material flow under all circumstances. An enclosure together with active dust suppression methods is the only possibility to minimise dust emissions.

In this case, a single DEM simulation might not necessarily assist in the design and the prediction of the material and air flow. However, the following approaches might be helpful to simulate and predict the air flow and the generated dust emission and hence help to improve the active dust suppression equipment:

- multiphase CFD simulation as documented in [66, 67],
- combined DEM and CFD simulation as reported in [68],
- coupled DEM–CFD simulation published in [69–72].

Coupled DEM–CFD simulations allow the possibility to visualise the development of the dust concentration, and hence, a direct comparison to experimental tests as shown in [72]. Here, a transfer chute with different enclosures at the receiving conveyor were experimentally tested and simulated. A good correlation between DEM–CFD simulation and experiment could be reached. However, such simulations are still an interesting and challenging research field due to the increased computational effort and the reduced possibility in the idealisation of the simulation model.

2.12 Interrelationships Between Wear, Build-Up, and Flow

As already mentioned previously in Section 2.2, the optimisation of transfer chute designs regarding the minimising of blockage risk and the minimising of wear can be contradictory. Hence, a balance is necessary.

While in some instances, the incline of the chute is sufficient to prevent build-up; for a fast-moving stream, this can also result in high wear, both of the chute and receiving belt. A typical steep incline chute loading point is illustrated in Figure 2.27 for a horizontal receiving belt.

Referring to Figure 2.28 and following on from Eqs. (2.9) and (2.10) presented earlier, the belt abrasive wear indicator, AWI can be expressed as:

$$\mathrm{AWI} = \left[(v_{\mathrm{Ratio}}\,X)^2\,(1 - v_{\mathrm{Ratio}}\,Y)\right] \quad (2.16)$$

where $v_{\mathrm{Ratio}} = v_{\mathrm{Exit}}/v_{\mathrm{Belt}}$; $\mathrm{X} = \sin(\psi + \alpha)$; $\mathrm{Y} = \cos(\psi + \alpha)$; ψ = chute incline; α = belt incline.

Using Eq. (2.16) above, the abrasive wear of the belt reaches a maximum when the loading velocity, v_{Loading}, is in the range of 50–90% of v_{Belt} [1]. Similarly, an increase in both the chute and belt inclination results in increased wear of the belt. This suggests that with steep chutes, v_{Loading} must be equal or very close to v_{Belt}, which is generally rarely achievable. Further, this also suggests that a much lower v_{Loading}, 40–50% of v_{Belt}, or even lower, is necessary to minimise wear on the belt. As an example, to minimise wear, for a chute inclination of 70° (to the horizontal), v_{Exit} must be lower than v_{Belt}, or approximately $3 \times v_{\mathrm{Belt}}$ [1]. For these reasons, steep chutes exhibit higher wear on the receiving conveyor belt as illustrated in Figure 2.28. Compared to a chute inclined at 45°, the abrasive wear for chute inclinations of 60° and 70° is observed to increase 3–16 times, depending on the inclination of the receiving conveyor belt.

Generally, as v_{Loading} approaches v_{Belt}, v_{Normal} approaches a maximum. This is illustrated in Figure 2.29. Also shown is the generalised interrelationship between transfer chute and conveyor belt velocities, wear and flow. The figure demonstrates that it is not favourable to optimise steep chute designs to match belt speed. Lower wear

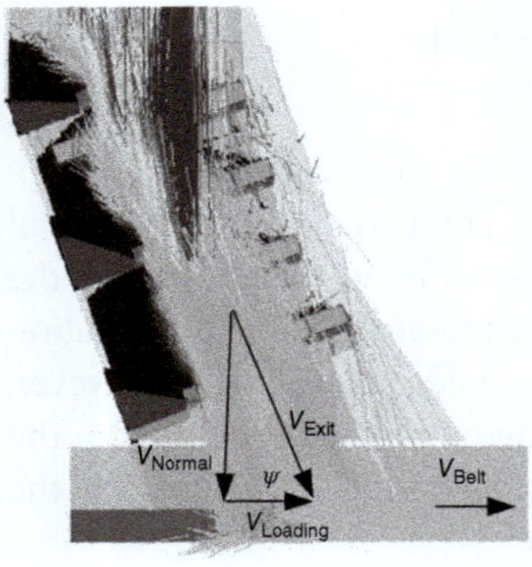

Figure 2.27 Steep loading chute.

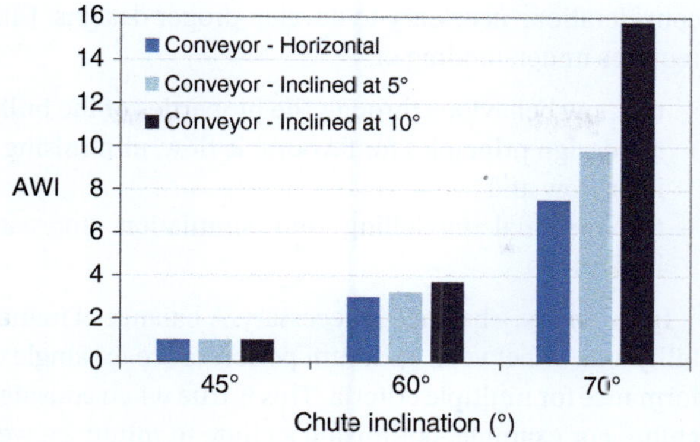

Figure 2.28 Relative abrasive wear (AWI) comparison of different chute and conveyor inclinations.

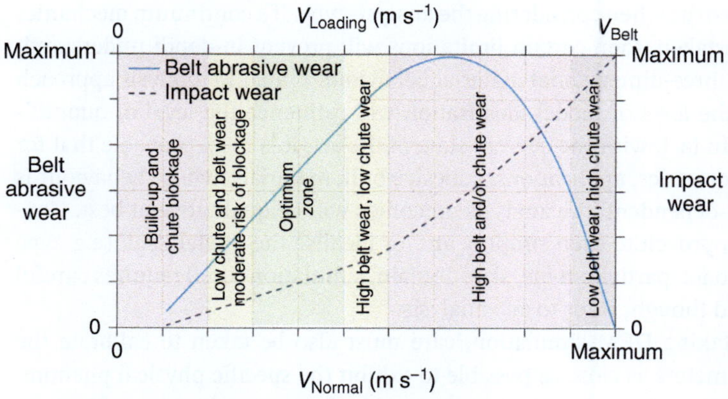

Figure 2.29 Transfer chute and belt velocity, wear, and flow relationships.

will generally occur at lower loading velocity, $v_{Loading}$; however, velocity must also be sufficient to maintain reliable flow and prevent blockage.

The relationship shown in Figure 2.29 also illustrates the balance of paradoxes and paradigms which are required to optimise the performance of both the conveyor belt and the transfer chute. It becomes clear that optimisation of conveyor belt wear performance does not always align with optimising flow and/or wear through the chute. On the other hand, it is also evident that the design of the chute can be compromised such that the wear life of the belt is maximised. Hence, a balance in objectives and therefore design criteria is necessary. To achieve this balance, performance-based criteria should be introduced as early as possible in the design process.

2.13 Conclusion

Transfer chutes are essential components of belt conveyor systems and materials handling facilities. In this chapter, we have presented a comprehensive overview of research and academic practice up to date that includes key elements and considerations necessary to develop proper designs. Efficient transfer chute design requires understanding of:

- the flow behaviour through the properties of the bulk solid material handled,
- the design principles for favourable flow, minimising wear and build-up, centralising flow and
- the essential modelling and simulation approaches, their strengths and limitations.

In summary, a balance is necessary. A balance of managing expectation with capability and/or between optimum performance for single criteria with acceptable performance for multiple criteria. This is true when considering the actual design of the chute. For example, optimising a chute to minimise wear may not necessarily also minimise build-up and may actually increase the propensity for blockage and vice versa. This is also true when considering the analysis type. If a continuum mechanics analysis is undertaken, then certain limitations will prevent in-depth understanding of complex, three-dimensional material behaviour. If a DEM analysis approach is undertaken, the level of model idealisation will influence the level of quantifiable outcomes. In fact, with the current state of the art, it is most probable that for real complex geometries, applications, cohesive bulk materials, whose behaviour is very much time-dependent, the analysis outcomes will be qualitative at best. Consequently, the approach to even simplify and/or idealise the model itself (e.g. type of simulation model, particle shape, size, domain, simulation time) requires careful investigation and thought prior to any analysis.

Before undertaking DEM simulation, care must also be taken to calibrate the simulation parameters as close as possible to exhibit the specific physical phenomena being investigated (i.e. localised regions of high wear or build-up, blockage with a specific set of simulation parameters and/or operating conditions). The chapter also highlights that the level of calibration and scientific interrogation to obtain quantitative DEM analysis outcomes such as wear life prediction needs to be significant and thorough.

Nomenclature

AWI	abrasive wear index
a	acceleration
A	cross-sectional area
A_0	initial cross-sectional area
α	conveyor inclination (to horizontal) or angle of impact
α_{t_0}	initial impact angle at the beginning of the contact at the time t_0
B	chute or stream width
C	constant
δ	effective angle of internal friction
ρ	bulk density
E_{Archard}	sliding energy according to Archard's model
E_{Normal}	deformation energy of impacting material stream

E_{Finnie}	cutting energy according to Finnie's model
ε	inclination of belt transition
F_D	drag force
F_n	normal force at contact point
g	gravitational acceleration
H	height of material burden on belt
H_0	initial height of stream
$k_{Archard}$	coefficient of sliding wear
k_{Finnie}	coefficient of cutting impact wear
k_{Normal}	coefficient of deformation impact wear
K	ratio of normal to tangential force approximated reciprocal of friction coefficient
K_V	normal to horizontal pressure ratio
μ	coefficient of sliding friction
μ_E	coefficient of equivalent friction
m_P	mass of particle
n	velocity exponent ranging from 1 … 1.5
N_{WR}	non-dimensional abrasive wear parameter
P	impact pressure
R	head pulley radius
φ	chute wall friction angle
s	sliding distance
Δt	time step
θ	chute inclination (to vertical)
v	velocity
v_0	initial velocity
v_{Belt}	belt velocity
v_{Exit}	velocity at chute exit
$v_{Loading}$	velocity component in the direction of conveyor belt travel
v_n	relative normal velocity at contact point
v_{Normal}	velocity component in normal direction
v_P	relative particle velocity at contact point
v_{Ratio}	ratio of velocity at loading
v_t	relative tangential velocity at contact point
ψ	chute incline (to horizontal)
w	wear parameter
W_a	abrasive wear parameter
x	horizontal co-ordinate
y	vertical co-ordinate

References

1 Ilic, D. (2019). Development of design criteria for reducing wear in iron ore transfer chutes. *Wear* 434–435: 202986.

2 Roberts, A.W. (1969). An Investigation into the gravity flow of non-cohesive granular materials through discharge chutes. *ASME Journal of Engineering for Industry* 91 (2): 373–381.

3 Donohue, T.J., Ilic, D., Bell, R., and Newman, L. (2010). The use of DEM in the design and analysis of WEARBACK transfer chutes. *Australian Bulk Handling Review* 11/12: 62–67.

4 Ilic, D., McBride, W., and Katterfeld, A. (2007). Validation of continuum methods utilising discrete element simulations as applied to a slewing stacker transfer chute. In: *9th International Conference on Bulk Materials Storage, Handling and Transportation, ICBMH 2007*, 11p.

5 Ilic, D. and Wheeler, C. (2017). Transverse bulk solid behaviour during discharge from troughed belt conveyors. *Advanced Powder Technology* 28 (9): 2410–2430.

6 Ilic, D., Roberts, A., and Wheeler, C. (2019). Modelling bulk solid interactions in transfer chutes: accelerated flow. *Chemical Engineering Science* 209: 115197.

7 Ilic, D., Roberts, A., Wheeler, C., and Katterfeld, A. (2019b). Modelling bulk solid flow interactions in transfer chutes: shearing flow. *Powder Technology* 354: 30–44.

8 Cundall, P.A. and Strack, O.D.L. (1979). A discrete numerical model for granular assemblies. *Geotechnique* 29: 47–65.

9 Ilic, D., Lavrinec, A., and Orozovic, O. (2020). Simulation and analysis of blending in a conveyor transfer system. *Minerals Engineering* 157: 106575.

10 Magalhães, M.F., Chieregati, A.C., Ilic, D. et al. (2021). Use of discrete element modelling to evaluate the parameters of the sampling theory in the feed grade sampler of a sulphide gold plant. *Minerals* 11 (9): 978.

11 Oldal, I. and Safranyik, F. (2015). Extension of silo discharge model based on discrete element method. *Journal of Mechanical Science and Technology* 29: 3789–3796.

12 Necas, J., Hlosta, J., Zurovec, D. et al. (2017). Feeder type optimisation for the plain flow discharge process of an underground hopper by discrete element modelling. *Advances in Science and Technology. Research Journal* 11: 246–252.

13 Otto, H. and Katterfeld, A. (2015). Prediction and simulation of the mistracking of conveyor belts. *8th International Conference on Conveying and Handling of Particulate Solids (CHOPS 2015), Tel-Aviv.*

14 Shen, J., Wheeler, C., Ilic, D., and Chen, J. (2019). Application of open source FEM and DEM simulations for dynamic belt deflection modelling. *Powder Technology* 357: 171–185.

15 Shen, J., Wheeler, C., O'Shea, J., and Ilic, D. (2018). Investigation of the dynamic deflection of conveyor belts via experimental and modelling methods. *Measurement: Journal of the International Measurement Confederation* 127: 210–220.

16 Žídek, M., Rozbroj, J., Jezerska, L. et al. (2021). Effective use of DEM to design chain conveyor geometry. *Chemical Engineering Research and Design* 167: 25–36.

17 Rossow, J. and Coetzee, C.J. (2021). Discrete element modelling of a chevron patterned conveyor belt and a transfer chute. *Powder Technology* 391: 77–96.

18 Davoodi, A., Asbjörnsson, G., Hulthén, E., and Evertsson, M. (2019). Application of the discrete element method to study the effects of stream characteristics on screening performance. *Minerals* 9 (12): 788.

19 (2005). *Belt Conveyors for Bulk Materials*, 6e. Conveyor Equipment Manufacturers Association (CEMA).
20 Roberts, A.W., Wiche, S.J., Ilic, D.D., and Plint, S.R. (2004). Flow dynamics and wear considerations in transfer chute design. In: *Proceedings of the 8th ICBMH 2004*Wollongong, Australia, 330–334. The Institution of Engineers.
21 Booth, E.P.O. (1934). Trajectories from conveyors – method of calculating them corrected. *Engineering and Mining Journal* 135 (12): 552–554.
22 Golka, K. (1992). Discharge trajectories of bulk solids. In: *International Conference on Bulk Materials Handling and Transportation, Symposium on Freight Pipelines, Conference Proceedings*, 497–503. Barton, ACT, Australia: Institution of Engineers.
23 Hastie D. (2010). Belt conveyor transfers: quantifying and modelling mechanisms of particle flow. PhD thesis. University of Wollongong.
24 Ilic, D., Wheeler, C.A., and Roberts, A.W. (2007). Bulk solid interactions in belt conveying systems: transition zone. *Particle and Particle Systems Characterization* 24 (4-5): 345–351.
25 Roberts, A.W. (2001). Chute design considerations for feeding and transfer. In: *Proceedings of the BeltCon 11 Conference*, Randburg, Republic of South Africa.
26 Roberts, A.W. (2003). Chute performance and design for rapid flow conditions. *Chemical Engineering & Technology* 26 (2): 163–170.
27 Roberts, A.W. and Scott, O.J. (1981). Flow of bulk solids through transfer chutes of variable geometry and profile. *Bulk Solids Handling* 1 (4): 715–727.
28 Hustrulid, A.I. and Mustoe, G.W. (1996). Engineering analysis of transfer points using discrete element analysis, Chapter 2. In: *Proceedings of the Annual Meeting of the Society of Mining Engineers (SME)*, Phoenix, AZ, 9–13.
29 Hustrulid, A.I. (1998). Transfer station analysis. In: *Proceedings of the Annual Meeting of the Society of Mining Engineers (SME)*, Orlando, FL, 33–53.
30 Katterfeld, A., Gröger, T., and Minkin, A. (2007). Discrete element simulation of transfer stations and their verification. *Bulk Solids and Powder Science Technology* 2 (2): 137–143.
31 Nordell, L.K. (1994). Palabora installs curved transfer chute in hard rock to minimize belt cover wear. *Bulk Solids Handling* 14 (4): 739–743.
32 Pinson, D., Reed, J., Wright, B., and Yu, A.-B. (2004). Application of discrete element particle simulation to flow in a transfer chute. In: *8th ICBMH 2004*, 294–298. University of Wollongong.
33 Cheng, H., Shuku, T., Thoeni, K. et al. (2019). An iterative Bayesian filtering framework for fast and automated calibration of DEM models. *Computer Methods in Applied Mechanics and Engineering* 350: 268–294.
34 Coetzee, C.J. (2017). Review: calibration of the discrete element method. *Powder Technology* 310: 104–142.
35 Mohsin, A., Roessler, T., Richter, C., and Katterfeld, A. (2020). Calibration of cohesive DEM parameters under rapid flow conditions and low consolidation stresses. *Powder Technology* 374: 22–32.
36 Roessler, T., Richter, C., Katterfeld, A., and Will, F. (2019). Development of a standard calibration procedure for the DEM parameters of cohesionless bulk

materials – Part I: solving the problem of ambiguous parameter combinations. *Powder Technology* 343: 803–812.

37 Doroszuk, B., Król, R., and Wajs, J. (2021). Simple design solution for harsh operating conditions: redesign of conveyor transfer station with reverse engineering and DEM simulations. *Energies* 14 (13): 4008.

38 Otto, H. (2019). Analyse und Simulation des Gurtschieflaufs an Gurtförderanlagen. PhD thesis. Otto-von-Guericke-Universität Magdeburg.

39 Chung, Y.C. and Ooi, J. (2008). Influence of discrete element model parameters on bulk behavior of a granular solid under confined compression. *Particulate Science and Technology* 26 (1): 83–96.

40 Do, H.Q., Aragaon, A.M., and Schott, D.L. (2017). Automated discrete element method calibration using genetic and optimization algorithms. In: *Powders and Grains 2017*, EPJ Web of Conferences, vol. 140, 15011.

41 Grima, A.P. and Wypych, P.W. (2011). Discrete element simulations of granular pile formation: method for calibrating discrete element models. *Engineering Computations* 28 (3): 314–339.

42 Rackl, M. and Hanley, K.J. (2017). A methodical calibration procedure for discrete element models. *Powder Technology* 307: 73–83.

43 Richter, C., Roessler, T., Katterfeld, A., and Will, F. (2020a). Development of a standard calibration procedure for the DEM parameters of cohesionless bulk materials – part II – efficient optimization-based calibration. *Powder Technology* 360: 967–976.

44 Roessler, T. and Katterfeld, A. (2019). DEM parameter calibration of cohesive bulk materials using a simple angle of repose test. *Particuology* 45: 105–115.

45 Dratt, M. and Katterfeld, A. (2017). Coupling of FEM and DEM simulations to consider dynamic deformations under particle load. *Granular Matter* 19 (3): 49. http://dx.doi.org/10.1007/s10035-017-0728-3.

46 Richter, C., Rößler, T., Otto, H., and Katterfeld, A. (2021). Coupled discrete element and multibody simulation, part I - implementation, verification and validation. *Powder Technology* 379: 494–504.

47 Engineerdo: Reasons for belt mistracking (2021). https://www.engineerdo.com/2019/10/04/belt-mistracking/.

48 Selby, M. (2010). Introduction to tracking a conveyor belt. *Australian Bulk Solid Handling.*

49 Katterfeld, A., Roberts, A., Wheeler, C. et al. (2021). Conveying and construction machinery. In: *Springer Handbook of Mechanical Engineering*, Springer Handbooks (ed. K.H. Grote and H. Hefazi), 829–991. Cham: Springer.

50 Ilic, D. and Donohue, T. (2012). Transfer chute analysis – current practices involving continuum, DEM and practical validation. *Australian Bulk Handling Review* 1/2.

51 Carr, M., Roessler, T., Otto, H. et al. (2019). Calibration procedure of Discrete Element Method (DEM) – parameters for cohesive bulk materials. In: *Conference proceedings ICBMH 2019*. Gold Coast: Institution of Engineers Australia.

52 Carr, M.J., Chen, W., Williams, K., and Katterfeld, A. (2016). Comparative investigation on modelling wet and sticky material behaviours with a simplified JKR cohesion model and liquid bridging cohesion model in DEM. In: *ICBMH2016 Barton*, Australia. A.C.T. Engineers.

53 Carr, M. (2019). Identification, characterisation and modelling of dynamic adhesion for optimised transfer system design. PhD thesis. The University of Newcastle.

54 Katterfeld, A., Coetzee, C.J., Donohue, T., et al. (2019). White Paper: Calibration of DEM Parameters for Cohesionless Bulk Materials under Rapid Flow Conditions and Low Consolidation. https://doi.org/10.13140/RG.2.2.26318.31048/1.

55 Ilic, D. and Hicks, T.J. (2016). Towards the development of design criteria for reduced wear in iron ore transfers. *17th Iron Ore Conference, ABM Week 2016*, Rio De Janeiro, Brazil, (September 2016).

56 Roessler, T. and Katterfeld, A. (2019). On the calibration of DEM parameters for abrasive sliding wear. In: *Proceedings of the 13th ICBMH*, Australia. Institution of Engineers.

57 Kloss, C., Goniva, C., and Katterfeld, A. (2012). Simulation of wear and dust emission at a transfer chute. *Cement International* 10 (3): 56–63.

58 Roessler, T. and Katterfeld, A. (2021). Untersuchungen zur Vorhersage des abrasiven Prall- und Gleitverschleiß auf Basis der Diskrete Elemente Methode. In: *Tagungsband 25. Fachtagung Schüttgutfördertechnik*. Otto-von-Guericke-Universität Magdeburg.

59 Chen, G. (2017). Surface wear reduction of bulk solids handling equipment using bionic design. PhD thesis. Delft University of Technology.

60 Wiche, S.J., Keys, S., and Roberts, A.W. (2005). Abrasion wear tester for bulk solids handling applications. *Wear* 258: 251–257.

61 de Oliveira, C., O Shea, J., Guo, J. et al. (2019). Investigation of impingement angle influence on impact wear of different chute liners by employing a novel impact wear tester. In: *13th International Conference on Bulk Materials Storage, Handling and Transportation ICBMH 2019*. Gold Coast.

62 Roessler, T., Katterfeld, A., Dratt, M. et al. (2020). Einsatz der Diskrete Elemente Methode zur Beschreibung des abrasiven Verschleißes in der Schüttgutfördertechnik. In: *Tagungsband 9. Kolloquium Fördertechnik im Bergbau*, 193–207. Institut für Bergbau, Technische Universität Clausthal.

63 Zhao, Y., Ma, H., Xu, L., and Zheng, J. (2017). An erosion model for the discrete element method. *Particuology* 34: 81–88.

64 VDI 3790 (2010). Emission of gases, odours and dusts form diffuse sources: storage, transhipment and transportation of bulk materials.

65 Bach, S. and Schmidt, E. (2008). Determining the dustiness of powders – a comparison of three measuring devices. *The Annals of Occupational Hygiene* 52: 717–725. https://doi.org/10.1093/annhyg/men062.

66 Donohue, T.J., Wheeler, C.A., Chen, X.L., and Katterfeld, A. (2012). Methods for the characterisation and control of dust in materials handling applications with a specific focus on passive dust control in transfer chutes. In: *Dust - Sources, Environmental Concerns, and Control: Sources, Environmental Concerns, and Control* (ed. L.T. Cacioppo), 99–142. New York: Nova Science Publishers.

67 Chen, X.L., Wheeler, C.A., Donohue, T.J. et al. (2012). Evaluation of dust emissions from conveyor transfer chutes using experiments and CFD simulation. *International Journal of Mineral Processing* 110, 111: 101–108.

68 Katterfeld, A., Donohue, T.J., and Wheeler, C.A. (2010). Transfer chute design: simulation-based dust prediction. *Bulk Solids Handling* 30 (5): 240–245.

69 Schulz, D., Schwindt, N., Schmidt, E., Jasevicius, R., & Kruggel-Emden, H.: Investigation of the dust release from bulk material undergoing various mechanical processes using a coupled DEM/CFD approach. *Powder Technology*, 355, 2019 pp. 37–56. https://doi.org/10.1016/j.powtec.2019.07.005.

70 Schulz, D., Schwindt, N., Schmidt, E., and Kruggel-Emden, H. (2021). Modelling of dust emissions induced by flow over stockpiles and through packed beds. *Particuology* 59: 55–63. https://doi.org/10.1016/j.partic.2020.10.014.

71 Hilton, J.E. and Cleary, P.W. (2013). Dust modelling using a combined CFD and discrete element formulation. *International Journal for Numerical Methods in Fluids* 72 (5): 528–549. https://doi.org/10.1002/fld.3750.

72 Goniva, C., Kloss, C., Chen, X. et al. (2014). Transfer chutes: predicting dust emissions by multiphase CFD and coupled DEM-CFD simulations. *Bulk Solids Handling* 34: 55–57.

3

Belt Conveyor Design and Troubleshooting

Craig A. Wheeler[1], Peter W. Robinson[1], Paul J. Munzenberger[1], Jiahe Shen[1,2], Michael J. Carr[1], Jayne I. O'Shea[3], and Shaun C. Reid[3]

[1] *The University of Newcastle, School of Engineering, Australia*
[2] *Wuhan University of Technology, School of Transportation and Logistics Engineering, China*
[3] *The University of Newcastle, TUNRA Bulk Solids, Australia*

Abbreviations

CFD	Computational fluid dynamics
DEM	Discrete element method
FDM	Finite difference method
FEM	Finite element method

3.1 Introduction

Belt conveyors are used extensively across the mining, quarry, processing, and agriculture industries to transport ore, minerals, rock, stone, and grain within plants and overland. Belt conveyors range in length from tens of metres to tens of kilometres and transport hundreds to thousands of tonnes of bulk material per hour. Belt conveyors are used in almost all large-scale bulk material handling and process plants and play a critical role across a broad range of industries. Their reliability and efficiency are critical, since in many cases they are the single means to supply raw bulk material to the plant and transport product from the plant.

Belt conveyor systems represent multibody systems comprising highly flexible belts, dynamic support elements, and the conveyed bulk material, leading to a complex dynamic system with varying physical properties that are highly dependent on the fundamental system design and operating conditions. The static and dynamic modelling of belt conveyors is an important step in the design process to ensure efficient and reliable systems, while numerical simulations are also often used to troubleshoot systems and provide greater insight into operational and maintenance issues.

Simulations in Bulk Solids Handling: Applications of DEM and other Methods, First Edition.
Edited by Don McGlinchey.

The state of art for current belt conveyor system design typically relies on a combination of specialised design software to model the dynamic characteristics of the system during starting, steady-state running, and stopping, and typically employ either the finite element method (FEM) or finite difference method (FDM) to model the elastic behaviour of the belt. The discrete element method (DEM) is typically employed to analyse the bulk material behaviour during loading, transport, and discharge and, in most cases, is implemented independent to the dynamic analysis of the conveyor belt. In some specialised cases, coupled FEM–DEM or FDM–DEM approaches are adopted to model and analyse particular system characteristics such as energy loss, belt wear, bulk material degradation and segregation, and bulk material stability. These analysis methods often require multiple and quite different software codes to interact with each other, and frequently require additional codes to transfer data.

This chapter will provide an overview of a range of simulation methods adopted to design and troubleshoot belt conveyors. The application of simulation methods to static belt conveyor design will be introduced by examining the complex interactions between the conveyor belt and idler rolls using the FEM, and the interaction between the conveyor belt and bulk material using coupled FEM–DEM. The FEM will be applied to verify critical stresses in conveyor pulleys and to predict the transient tensions induced in the conveyor belt during starting and stopping. The DEM will be employed to model the bulk material flow during loading and discharge, and assist in troubleshooting problematic systems, while CFD will be applied to model air flow and entrained dust to predict spillage.

3.2 Belt Conveyor Design

This section will present a range of simulation methods and approaches to predict the starting, steady-state running, and stopping of belt conveyors in addition to the application to component design.

3.2.1 Static Analysis Design Methods

During the design of a belt conveyor system, the demand power, take-up tension, and resulting steady-state belt tension distribution are calculated for a range of loading conditions. Derivation of the various resistance components varies according to the standard or design method followed, while slope resistances calculations generally follow the same procedure. For the purpose of this chapter, a straight horizontal belt conveyor is assumed. Figure 3.1 outlines the components of motion resistance for a horizontal belt conveyor. The motion resistance components include indentation rolling resistance (IRR), belt flexure resistance, idler roll rotating resistance, idler roll misalignment resistance, and bulk material flexure resistance. The cumulative resistance along the length of the belt conveyor is calculated by superimposing the contributions from each of the motion resistances, culminating in an effective belt tension, T_e, due to the sum of the resistances. T_1 represents the maximum belt

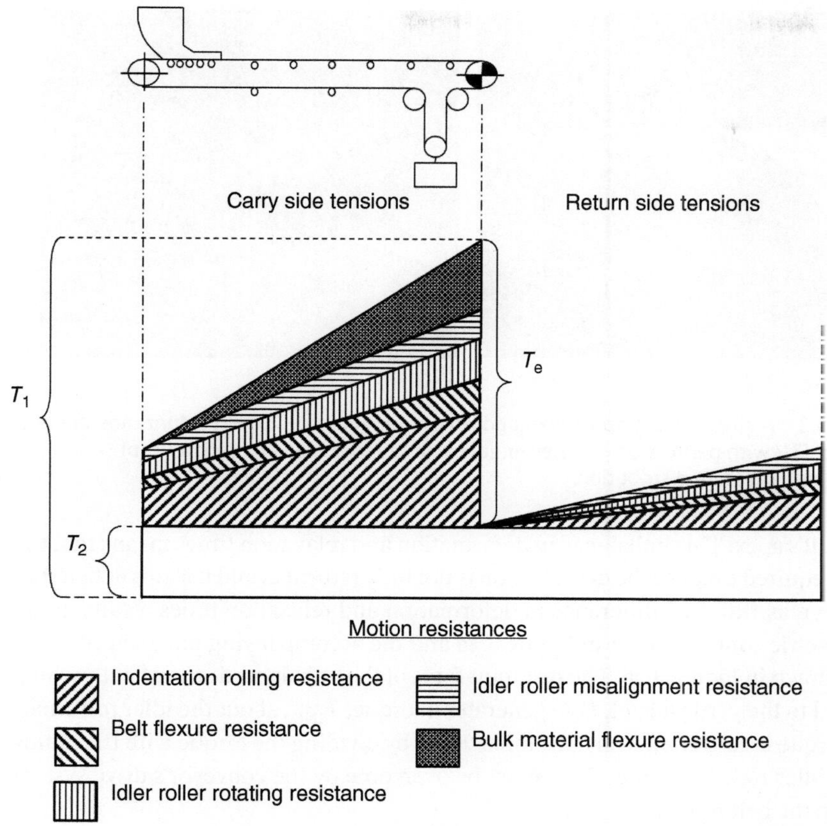

Figure 3.1 Steady-state belt tension distribution for a horizontal belt conveyor.

tension at the drive pulley, while T_2 represents the necessary pre-tension to ensure the drive pulley is able to transmit the effective belt tension, T_e, without slip.

With reference to Figure 3.1, the belt and bulk material flexure resistance components are shown (due to scaling) to increase linearly with length and therefore are independent of belt tension. In practice, these components do in fact decrease per unit length, as belt tension increases. All other resistance components are assumed independent of belt tension in the absence of vertical and horizontal curves.

The IRR, and belt and bulk material flexure resistance are discussed, and numerical models presented to show the basis for predicting each resistance component. For the purpose of this chapter, the idler roll rotating resistance and idler roll misalignment resistance will not be discussed, with the reader referred to Wheeler [1, 2] for a comprehensive treatment of these components.

3.2.1.1 Numerical Analysis of Indentation Rolling Resistance

IRR is a friction-like force that is developed in a conveyor system by the interaction of the conveyor belt's pulley cover and the idler rolls that the conveyor belt travels upon. A section of conveyor belt that has been compressed or indented in a given time as it approaches the apex of an idler roll will take longer to relax as it leaves the

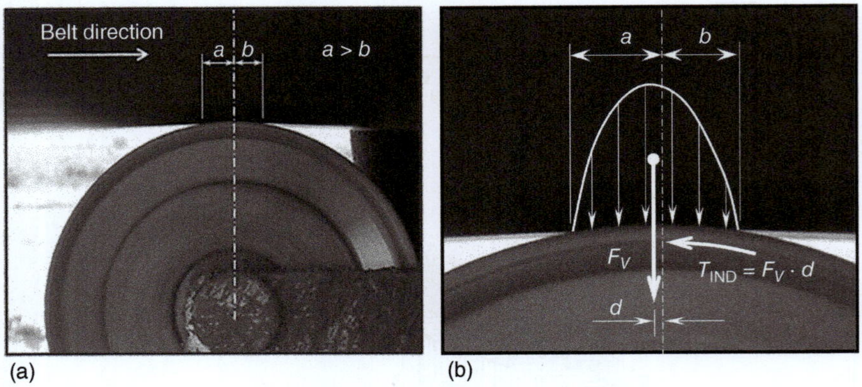

Figure 3.2 A stress field graph superimposed over an idler roll. Source: Munzenberger and Wheeler [3]/ with permission of, Elsevier. (a) Cyclic compression and recovery. (b) Asymmetric pressure distribution.

idler roll's apex. This imbalance in deformation and relaxation times means that the effort required to cause the deformation is not fully returned, and it is this deficit that is known as IRR. The difference in deformation and relaxation times results in the asymmetric contact shown in Figure 3.2a and the accompanying unbalanced stress field shown in Figure 3.2b. The resultant force of the unbalanced stress field – which is equal to the vertical load, F_V – generates a torque, T_{IND}, about the idler roll's axis. The torque is converted to an IRR force, F_{IND}, by dividing the torque with the radius of the idler roll. The force, F_{IND}, must be overcome by the conveyor's drive system to keep the belt moving.

3.2.1.2 Simulation of Indentation Rolling Resistance with the Finite Element Method

To predict IRR using the FEM, a technique known as *linear viscoelastic analysis* is used. The linear part of the analysis assumes that it is valid to sum to the current strain, experienced by a control volume during the current timestep with the residual strain effects from strains experienced at each past timestep. The analysis is similar to the typical FEM procedure where the local stiffness matrix $[k]$ is generated for each element and assembled into the global stiffness matrix $[K]$. The equation $[K]\{D\} = \{R\}$ is then solved to generate nodal displacements. The difference between the typical FEM and a linear viscoelastic finite element analysis is that there are additional elements, aside from the typical mesh elements, which link the strains of one element to the stiffness of another. These additional elements mimic the memory effect of viscoelastic materials.

With reference to Figure 3.3, the conveyor belt, modelled as moving towards the right, is discretised into strips. Each element in a strip represents all elements to its right at a specific timestep in the past. Considering control volume R at the current time step, the control volumes, K_n, represent R at each prior time step before the current time. Each Kth element is related to each R element by a viscoelastic memory element. The set of viscoelastic memory elements are then distributed throughout

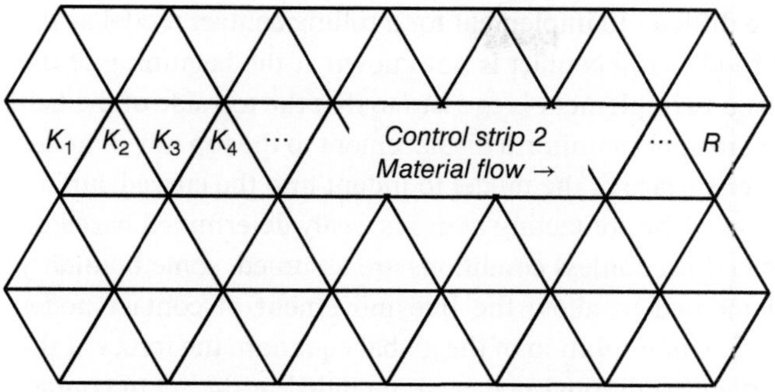

Figure 3.3 Diagram of a finite element analysis mesh for a viscoelastic rolling contact problem.

the global stiffness matrix of the model along with the mesh elements such that they relate the displacements of each K element to the nodes of element R.

For a plane strain mesh of three node constant strain triangle elements, the local stiffness matrices for the meshed elements and the viscoelastic memory elements are calculated with Eq. (3.1) in essentially the same manner as would be detailed in any finite element analysis test, Cook [4] for example, except that the elastic stiffness matrix is replaced with the matrix in Eq. (3.2).

$$[k]_{RK} = [B]_R{}^T[E]_{RK}[B]_RA_R \tag{3.1}$$

$$[E]_{RK} = \begin{bmatrix} a_{RK} & b_{RK} & 0 \\ b_{RK} & a_{RK} & 0 \\ 0 & 0 & \dfrac{a_{RK} - b_{RK}}{2} \end{bmatrix} \tag{3.2}$$

where:

$$a_{RK}\begin{cases} \frac{2}{3}G_1(1) + \frac{1}{3}G_2(1) & K = R = 1 \\ \frac{1}{2}\left(\left(\frac{2}{3}G_1(1) + \frac{1}{3}G_2(1)\right) + \left(\frac{2}{3}G_1(T_{R-1,R}) + \frac{1}{3}G_2(T_{R-1,R})\right)\right) & K = R > 1 \\ -\frac{1}{2}\left(\left(\frac{2}{3}G_1(T_{2,R}) + \frac{1}{3}G_2(T_{2,R})\right) - \left(\frac{2}{3}G_1(T_{1,R}) + \frac{1}{3}G_2(T_{1,R})\right)\right) & K = 1, R > 1 \\ -\frac{1}{2}\left(\left(\frac{2}{3}G_1(T_{K+1,R}) + \frac{1}{3}G_2(T_{K+1,R})\right) - \left(\frac{2}{3}G_1(T_{K-1,R}) + \frac{1}{3}G_2(T_{K-1,R})\right)\right) & 1 < K < R \end{cases} \tag{3.3}$$

And b_{RK} is obtained by replacing all $\frac{2}{3}G_1$ terms with $-\frac{1}{3}G_1$. G_1 and G_2 are the shear modulus and bulk modulus, respectively, and $(T_{a,\,b})$ represents the amount of time between control volumes a and b and is the time at which G_1 and G_2 are chosen from the shear and bulk modulus relaxation data. As each of the individual element stiffness matrices is computed, they are assembled into the global stiffness matrix. Care is required to distribute the local stiffness matrices of the viscoelastic memory elements such that they relate the displacements of the Kth element to the stiffness of the Rth element. This unusual distribution results in an asymmetric global stiffness matrix.

Boundary conditions are difficult to implement for a rolling contact model as the amount of conveyor belt to idler roll contact is not known at the beginning of the analysis. The easiest method to implement is to assume that the top side of the belt model is to remain flat and to apply a uniform displacement to the top row of nodes which act to force the lower surface of the model to indent into the curved surface of an idler roll. The position of the contacting nodes is easily determined based on the amount of indentation. If frictionless conditions are assumed, some boundary condition must be implemented to allow the free movement of contact nodes around the idler roll surface. Upon solution of the global equation, the forces at the contact nodes are assessed and additional nodes are included in the set of contact nodes until nodes with tensile contact reactions are found. The global equation is solved in each iteration. When contact nodes with tensile reactions are found, they are released, and a new solution is calculated. The final solution is determined when the set of contact nodes is the complete set of contact nodes with no contact nodes having tensile reactions.

When the correct set of contact nodes is found, the vertical reaction force at each contact node is found and multiplied by the nodes distance from the imaginary idler roll centre. The moments are then summed, and the result is divided by the radius of the conveyor idler to determine the IRR of the model. The vertical node reactions also need to be summed, without finding the moments, so that the belt load may be determined. If the sum of the loads is not equal to the required idler roll load, the indentation depth is modified, as appropriate, and the process is repeated. Rather than searching for a specific idler roll load, it is preferable to increment the indentation depth by some value and increment the depth until the desired maximum idler roll load is achieved. At each indentation depth, the IRR and idler roll load is recorded so that a trend can be developed.

With a set of idler roll loads, x, and their equivalent indentation rolling resistances, F_{IRR}, calculated, they are fitted to Eq. (3.4), by any convenient means, to determine a value for S which is used to apply the IRR modelling results to a troughed conveyor. In Eq. (3.4), $c = 4/3$.

$$F_{\text{IRR}} = Sx^c \tag{3.4}$$

Material Inputs Material property data for viscoelastic finite element model are time-dependent shear relaxation moduli and bulk moduli. A set of shear relaxation moduli for different strains are shown in Figure 3.4. The time-dependent bulk moduli are difficult to measure and have little bearing on the viscoelastic analysis. The bulk modulus may be approximated from the relaxation modulus by assuming a constant Poisson's ratio for rubber or be a constant of sufficiently large value. To measure the shear relaxation properties of rubber, a machine called a dynamic mechanical analyser like the one pictured in Figure 3.5a is used. A rubber specimen is clamped into the machine and is oscillated from nearly zero strain to the desired strain level. One possible clamping arrangement is shown in Figure 3.5b, and good results have also been measured with a tensile test and the appropriate conversions. While the specimen is cycled, the force and strain magnitude are continually

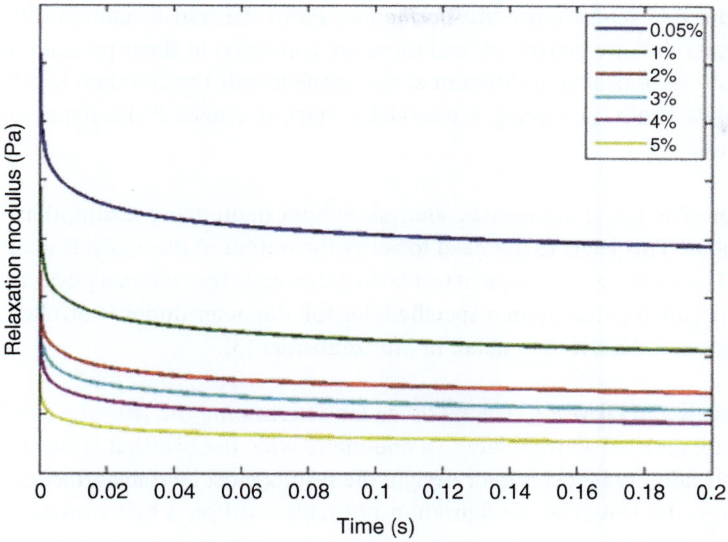

Figure 3.4 Shear relaxation moduli.

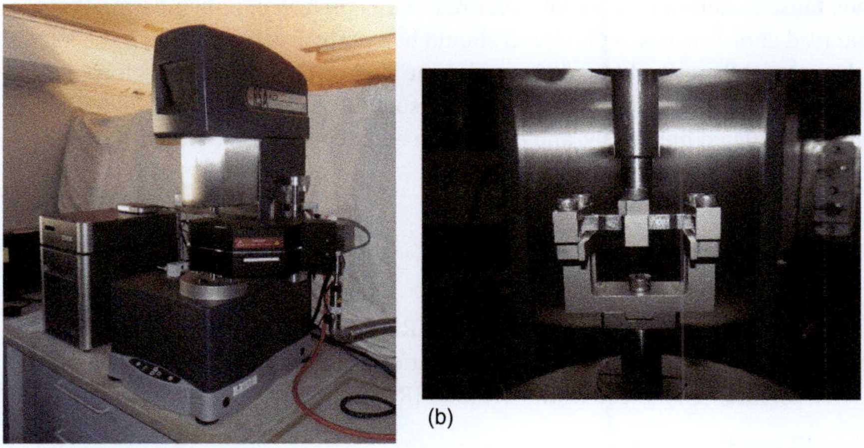

(a)

(b)

Figure 3.5 Measurement technique for determination of rubber relaxation properties. (a) RSA G2 Dynamic Mechanical Analyser for measuring rubber relaxation properties. (b) Dual shear rubber relaxation arrangement.

monitored so that the peak stress level and phase difference between them can be determined. The same test is repeated at a range of frequencies, and the phase and stress data from each test are used to calculate the storage and loss modulus for those conditions. Unfortunately, the dynamic mechanical analyser is not capable of measuring at the required frequencies, so an entire set of tests conducted at low frequencies is repeated at selected temperatures from −60 up to 60 °C, depending on the material, and low-frequency data are converted to high-frequency data through a process called a *time-temperature-transformation*. From the transformation, the storage and loss moduli are generated, and these are converted to shear relaxation moduli. The tests are repeated at different strain levels to suit the situation being modelled, with 2% strain data being a reasonable starting choice if the required values are unknow.

Model Calibration The linear viscoelastic analysis is built upon many assumptions and simplifications. This leads to the need to verify the output of the analysis with a physical test. The most accurate type of test is the large-scale test where a physical conveyor belt specimen is tested on a specific idler roll diameter under controlled conditions. This test is described in detail in Munzenberger [3].

Conversion of Results to a Conveyor The methods for determining the IRR of a conveyor belt that are detailed above relate to a one-metre-wide flat belt that is evenly loaded. For application to belt conveyor design, the results must be transformed so that they relate to the troughed configuration normally used on a belt conveyor. A three-roll trough arrangement is shown in Figure 3.6a along with an assumed normal reaction load distribution due to the weight of the conveyor belt – $q_{b,c}$ and $q_{b,s}$ – and the bulk material – $q_{m,c}$ and $q_{m,s}$ – being conveyed. Each of these reaction loads must be determined in whatever manner is preferred, remembering that the axial load above each wing idler roll should be evenly distributed across the length of the centre idler roll; although, in reality, the load will be carried at each end of the centre roll. The lengths $L_{m,s}$ and L_c must also be determined. These values, along with S, calculated previously, and $c = 4/3$ are entered into Eqs. (3.5) and (3.6).

$$F_{\text{IRR},c} = SL_c^{(1-c)}(q_{m,c} + q_{b,c})^c \tag{3.5}$$

$$F_{\text{IRR},s} = \frac{SL_{m,s}^{(1-c)}}{2q_{m,s}(c+1)}\left[(2q_{m,s} + q_{b,s})^{(c+1)} - q_{b,s}^{(c+1)}\right] \tag{3.6}$$

These equations calculate the sum of the IRR shown in Figure 3.5(b) for the centre and wing idlers. The individual indentation rolling resistance values are then entered into Eq. (3.7) to determine the total IRR for a single idler set.

$$F_{\text{IRR}} = F_{\text{IRR},c} + 2F_{\text{IRR},s} \tag{3.7}$$

3.2.1.3 Numerical Analysis of Belt and Bulk Solid Flexure Resistance

Flexure resistance occurs due to the cyclic longitudinal and transverse belt deflection between consecutive idler sets. It consists of belt flexure resistance and bulk solid flexure resistance. Belt flexure resistance occurs to the transverse and longitudinal bending of the belt and the resulting hysteresis losses, while bulk solid flexure is due to the internal friction of the bulk solid and interaction with the conveyor belt.

Bulk solid flexure resistance is generally considered to be the second greatest motion resistance in a belt conveyor system. Due to the complicated interaction of the belt and bulk solid, it is difficult to investigate in isolation. A number of theoretical models have been developed to analyse the interaction and related flexure resistance based on the Coulomb wedge theory or Rankine earth pressure

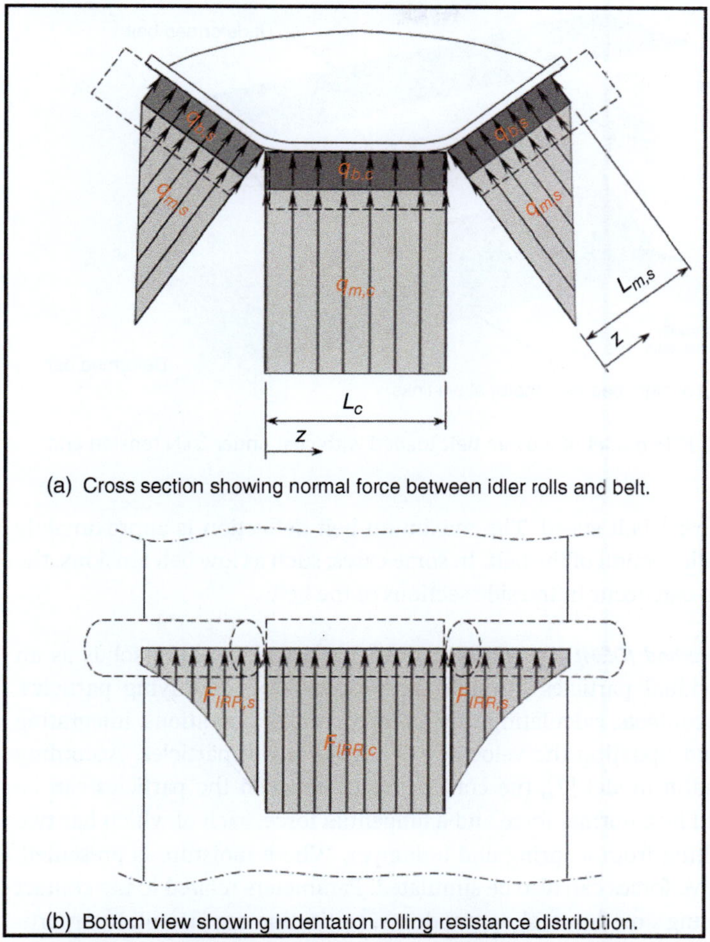

Figure 3.6 Approximate mass distribution (a) and indentation rolling resistance distribution (b) for a trough idler roll set. Source: Munzenberger and Wheeler [3]/with permission of Elsevier.

theory, such as those developed by Krause and Hettler [5], Spaans [6], and Wheeler [2]. This section introduces the use of the FEM and the DEM in modelling the deflection and interactions of conveyor belts.

Finite Element Method (FEM) The FEM commonly idealises conveyor belts as plate elements and models them using an orthotropic material with different properties, including Young's modulus, shear modulus, and Poisson's ratio. Element loads and boundary conditions are assigned to create the mathematical model. A non-linear static solver is generally adopted due to the geometry and material nonlinearities of conveyor belts. An example is shown in Figure 3.7 for a 6 mm thick belt supported by idler rolls at a 1200 mm spacing in the FEM software – Strand7. The belt is loaded with a bulk material with a bulk density of 850 kg m^{-3} under a 5 kN tension and

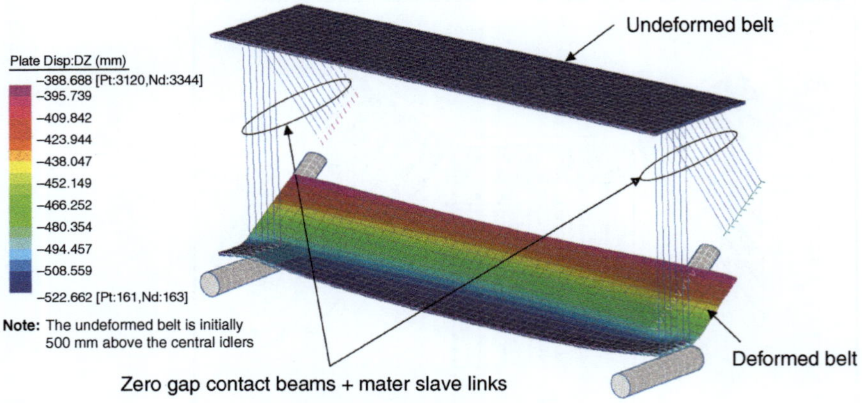

Figure 3.7 Strand7 FEM model of a 6 mm belt loaded with coal under 5 kN tension and $1\ \mathrm{m\,s^{-1}}$ belt speed.

simulated at a $1\ \mathrm{m\,s^{-1}}$ belt speed. The maximum belt deflection is approximately 22 mm in the middle section of the belt. In some cases, such as low belt tensions, the maximum belt sag can occur in the side sections of the belt.

Discrete Element Method (DEM) In contrast, the DEM considers bulk solids as an assembly of individual particles through the processes of identifying particles, detecting particle contacts, calculating contact forces and accelerations, integrating over timesteps, and updating the velocities and positions of particles. According to the Hertz–Mindlin model [7], the contact forces between the particles can be typically computed by a normal force and a tangential force, each of which has two components resulting from a spring and a damper. Where moisture is presented, adhesive or cohesive forces can also be simulated. Parameters related to the contact model include Young's modulus, shear modulus, restitution coefficient, interparticle friction, wall friction, and rolling friction. These parameters are required to be calibrated before modelling to improve the simulation accuracy. Figure 3.8 shows the same belt mentioned in "Finite Element Method (FEM)" section modelled by the DEM software - LIGGGHTS. From the DEM simulation, the particle velocities and loads applied on the belt can be derived.

FEM and DEM Coupling Simple engineering problems can be simulated via either the FEM or the DEM; however, when dealing with a belt conveyor that involves both continuum (the belt) and non-continuum (the particles), it is preferred to couple the FEM and DEM. Surface coupling is adopted for modelling as it is the most straightforward coupling method [8], and in this application, the belt is modelled in the FEM and the bulk solids in the DEM, and the data of the belt deformation and particle loads are exchanged. Depending on how the data are exchanged, the coupling can be classified as one-way or two-way. For one-way coupling, the data transfer occurs once, from the DEM to the FEM, as shown in Figure 3.9. Its main drawback is that it does not transfer the FEM mesh back to the DEM, resulting in the boundary conditions in the DEM being overlooked.

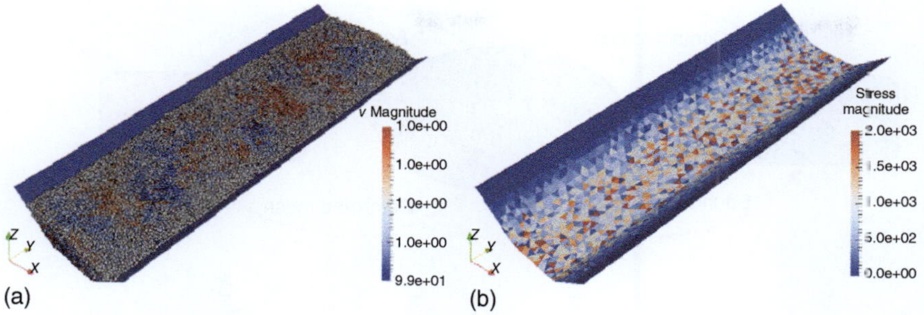

Figure 3.8 DEM belt conveyor model in LIGGGHTS. (a) DEM simulation showing particle velocities. (b) DEM simulation showing particle loads.

Figure 3.9 The FEM–DEM coupling scheme based on a Python programmed interface.

Two-way coupling exchanges the FEM and DEM data mutually in an iterative manner ensuring interdependence of the belt deflection and the load from the bulk solid is considered. The scheme and data flow of the two-way coupling are shown in Figure 3.10. In this case, the coupling happens every 1000 DEM timesteps, or every FEM timestep. The analysis will terminate when the change in belt deflection between two consecutive iterations reaches its predetermined convergence limit.

Figures 3.11 and 3.12 show simulations of a conveyor belt transporting iron ore, where Figure 3.11 shows the first coupling iteration, and Figure 3.12 shows the last iteration. The implementation of the FEM and DEM coupling is conducted in the open-source software Code_Aster and LIGGGHTS, respectively. Better modelling accuracy can be achieved when compared to one-way coupling as the DEM boundary conditions are periodically updated to simulate the deformed belt.

3.2.2 Dynamic Analysis Design Methods

Belt conveyors consist of multiple moving and stationary components distributed along the length of the system, as well as varying loads and drag forces. These, coupled with the viscoelastic nature of the belt itself, cause periods of acceleration of the belt to be non-instantaneous. Instead, when a conveyor belt is started or stopped, a transient wave propagates along the system, carrying with it the change in velocity. A simple analogy is that of a slinky under cyclic compression or tension.

The transient wave initiates at the source of the velocity change; typically, either a drive station or brake. In the case of head-drive starting, a tension wave propagates from the drive, along the carry side of the belt, with a corresponding compression wave propagating along the return side. These waves have the ability to interact with

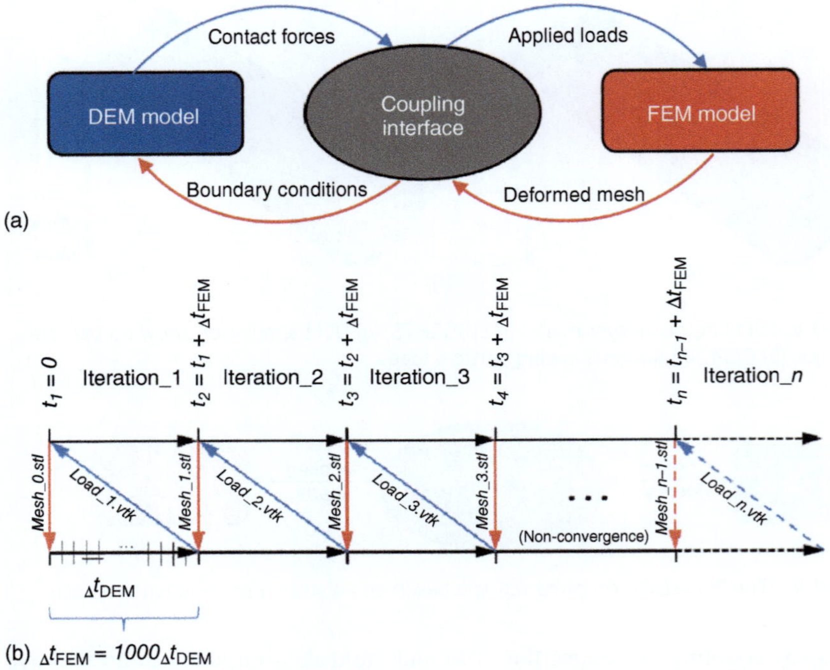

Figure 3.10 Two-way coupling. Source: Dratt and Katterfeld [9]/with permission of Springer. (a) Simulation scheme. (b) Data flow chart.

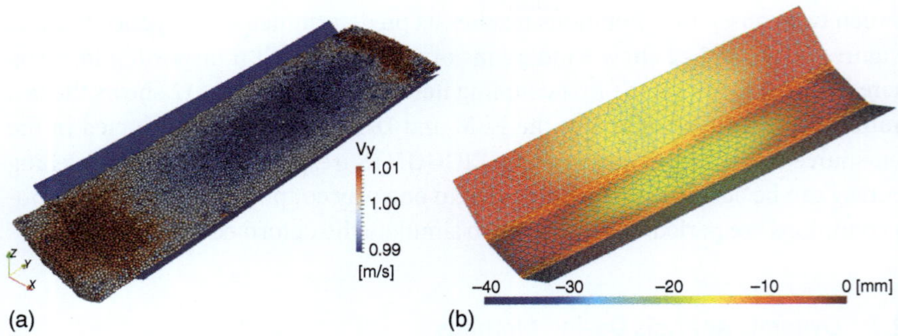

Figure 3.11 Two-way coupling at first iteration. (a) Particle velocities in LIGGGHTS. (b) Belt deformation in Code_Aster.

each other, or pieces of high mass infrastructure (e.g. another drive system, counterweight), resulting in either cancellation, reflection, or the formation of a standing wave with different behaviour. For small installations, this behaviour is less obvious due to the lower mass of the system, and as such, can be easily accounted for through an appropriate safety factor. For longer installations, however, or those with severe starting or stopping requirements, these dynamics play an increasingly important role in the optimum design and installation of a belt conveyor.

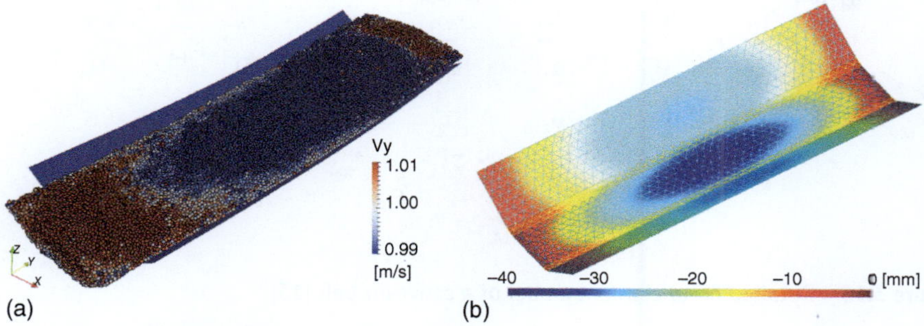

Figure 3.12 Two-way coupling at final iteration. (a) Particle velocities in LIGGGHTS. (b) Belt deformation in Code_Aster.

Current design methods such as ISO5048 [10] and DIN22101 [11] assess the static behaviour of a system operating at constant velocity to determine belt tension distribution and the required drive power. Some consideration is given to the dynamic behaviour of the system; however, these methods assume a rigid and inextensible configuration along the belt, which behaves according to Newtonian mechanics. These methods are acceptable for short installations, however, would result in an overly conservative and expensive approach for longer installations.

As the transient velocity wave represents a travelling region of high or low tension, several problems may arise if not correctly accounted for. The most common of which include:

- Failure of the belt or splice
- Excessive displacement of the counterweight system
- Excessive belt sag or tracking issues
- Belt lift-off in vertical curves, or
- Damage to supporting infrastructure.

3.2.2.1 Modelling of the Conveyor Belt Using Finite Elements

For conventional conveyor systems, multiple theories have been developed to estimate the dynamic belt behaviour during starting and stopping [12–15]. Initially, these modified the traditional wave equation to represent the stress variations that propagate as sound waves in a conveyor, while more recently the FEM has been applied.

The FEM breaks a conveyor system into a series of fundamental elements. Each element represents a given length of the system, typically in the order of 10–50 m, depending on the length of the system and desired accuracy. The system mass associated with this length is lumped together and connected by a series of spring and dashpot elements representative of the conveyor belt. An example FEM model of the conveyor system is shown in Figure 3.13, defining the position (x) of each element, the corresponding mass (M_1 or M_2), and the forces in between. The inter-nodal forces are governed by the elastic properties of the belt, and will vary with belt structure and conveyor design. An example model is given in Figure 3.14, depicting the belt carcass with stiffness K_0, and a 3-parameter Maxwell model to represent the

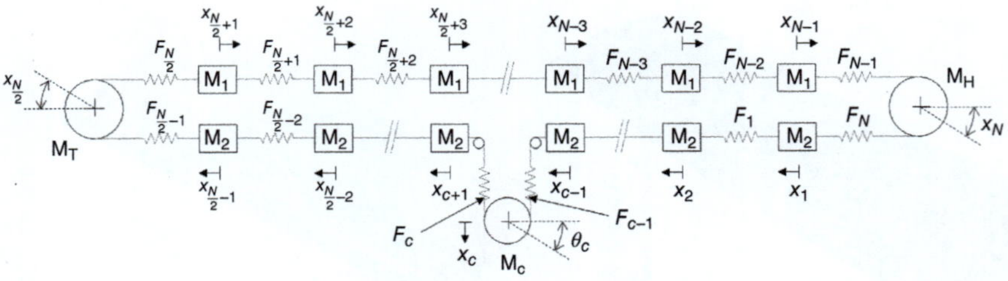

Figure 3.13 Detailed finite element model of a conveyor belt [15].

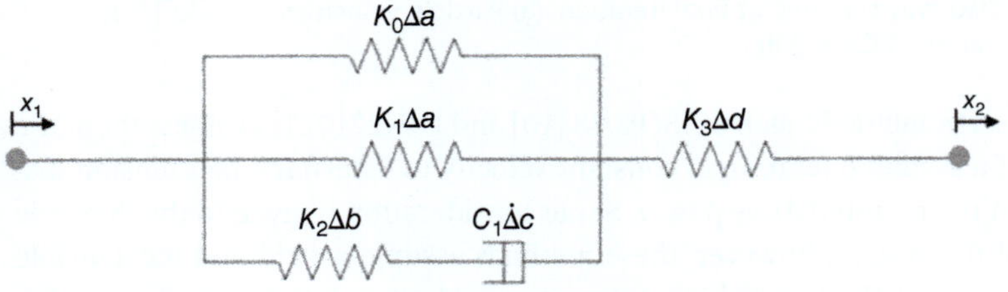

Figure 3.14 Mathematical representation of the finite element model [15].

rubber. These parameters are determined through tensile or cyclic testing of the conveyor carcass and rubber compounds. The term K_3 accounts for the variable stiffness along the length of the belt due to sag of the belt between the idlers [13].

For the dynamic model shown in Figure 3.13, the mass of each element contributes to both the inertia of the system and gravitational load based on a change in elevation. When considering the tension in the system due to gravity, the mass consists of the belt and bulk material load only. To determine the system mass when starting and stopping, the inertia of all rotating masses must be accounted for, including idler rolls, flywheels connected to the drive system, and pulleys. The mass of these elements is not necessarily the physical mass of the component, as it is dependent on the radius at which this mass is moving. The principle of reduced mass is employed to translate the inertia of components rotating at different radii to a single mass moving with linear velocity at the belt-line.

By definition, the moment of inertia, I, is given by:

$$I = \sum m_i R_i^2 \tag{3.8}$$

If the moment of inertia of a component is known, the reduced mass of all components acting at a radius of R is shown below. The value of R typically relates to the distance between the centre of rotation of the component and the belt line.

$$m_{\text{red}} = \frac{I}{R^2} \tag{3.9}$$

With the inertial mass known, the dynamic behaviour is dependent on the external forces acting on each element. These consist of the inter-nodal forces connecting each element, gravitational forces associated with the elevation of the static weight of the element, drag forces associated with conveyor operation, and any drive or

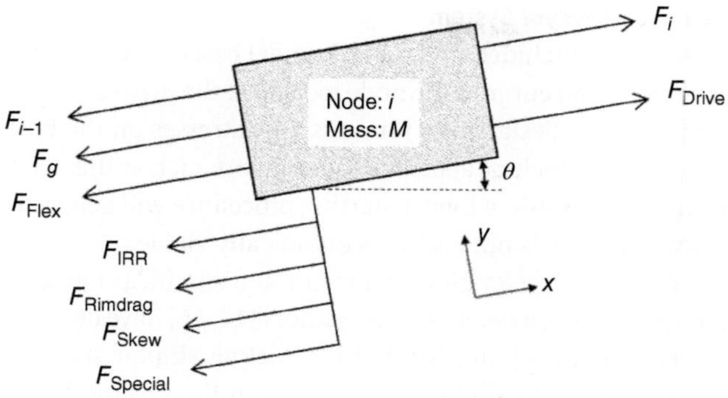

Figure 3.15 Free body diagram of a belt element [15].

braking force associated with the element location. A free-body diagram is shown below depicting the common forces included in an FEM model (Figure 3.15).

The primary benefit of an FEM model is the ability to localise various parameters along the conveyor length. This simplifies assessment of a varying material load on the belt, or changes to the idler pitch. Similarly, it is simple to interconnect parameters that may be related. For instance, it is well understood that various drag forces such as IRR and rim drag vary with belt speed, while other parameters such as belt and material flexure vary with belt tension. These inter-relationships can be included in real-time in an FEM to ensure analysis of the highest accuracy.

With known masses, forces, and elastic parameters, several options are available to solve the FEM model, including graphical simulation packages (e.g. Simulink), or alternatively, an Ordinary Differential Equation (ODE) can be derived to define the force between the masses, F_i, and solved in a computational software package. This process is shown below for the model shown in Figure 3.13. Defining the extension of an element as:

$$\Delta = x_2 - x_1 \tag{3.10}$$

The ODE defining the inter-nodal force of node i is given below:
where:

$$\dot{F}_i + JF_i + N\dot{\Delta} + P\Delta = 0 \tag{3.11}$$

$$J = \frac{K_2}{C_1}\left[\frac{K_0 + K_1 + K_3}{K_0 + K_1 + K_2 + K_3}\right] \tag{3.12}$$

$$N = -K_3\left[\frac{K_0 + K_1 + K_2}{K_0 + K_1 + K_2 + K_3}\right] \tag{3.13}$$

$$P = \frac{-K_2K_3}{C_1}\left[\frac{K_0 + K_1}{K_0 + K_1 + K_2 + K_3}\right] \tag{3.14}$$

These relationships are utilised with Newton's 2nd law of motion to determine the position, velocity, and tension of each element along the system.

3.2.2.2 Starting Sequences of a Conveyor System

Starting and stopping sequences are included in the FEM model based on either the prescription of a torque or acceleration curve to the node located at the drive or brake station. This curve will frequently be designed to minimise the stresses on the belt and conveyor infrastructure and may include specific requirements such as the need to stop within a certain distance. Naturally, a longer starting procedure will generate less strain on the system; however, this is not always economically viable.

A starting sequence with constant acceleration will result in a significant jerk at the beginning and end of the procedure; as such several authors [12, 16] have derived an S-shape start-up curve. Harrison [12] and Nordell [16] define similar starting characteristics that were discovered to exhibit minimal stress on the system. For a running belt velocity of v_b, and starting procedure of length T, Harrison prescribes a sinusoidal velocity profile, as given by:

$$v_{\text{Drive}}(t) = \frac{v_b}{2}\left(1 - \cos\left(\frac{\pi t}{T}\right)\right) \quad 0 \le t \le T \tag{3.15}$$

Similarly, Nordell specifies a parabolic start-up:

$$v_{\text{Drive}}(t) = v_b\left(\frac{2t^2}{T^2}\right) \text{ for } 0 \le t \le \frac{T}{2} \tag{3.16}$$

$$v_{\text{Drive}}(t) = v_b\left(-1 + \frac{4t}{T} - \frac{2t^2}{T^2}\right) \text{ for } \frac{T}{2} \le t \le T \tag{3.17}$$

These are compared below in Figure 3.16. Both profiles are similar, with Nordell exhibiting a slightly steeper acceleration.

In addition to these S-shape profiles, it is becoming increasingly common to utilise a low-velocity inspection period upon start-up. These periods are regions of constant low velocity, which perform two functions: the conveyor overcomes its break-away resistance, and visual inspections of belt can be performed, with emergency stopping from a low velocity easier if required. Lodewijks [13] states that the utilisation of an

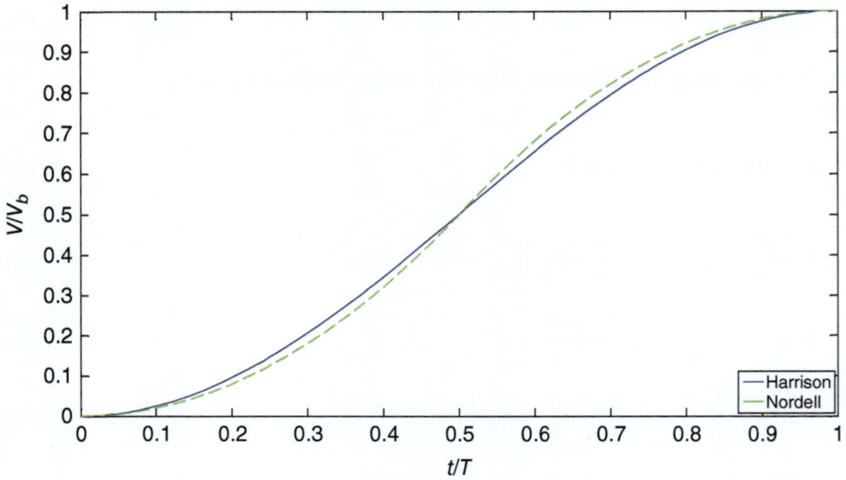

Figure 3.16 S-type starting procedures as given by Harrison [12] and Nordell [16].

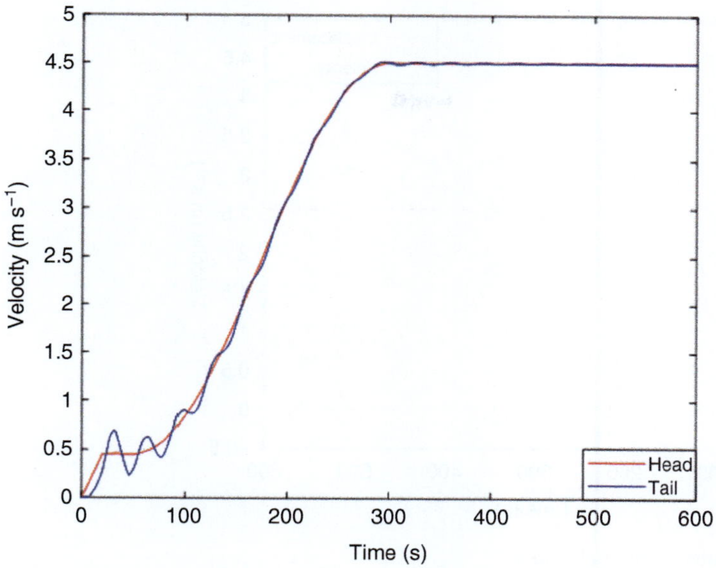

Figure 3.17 Belt velocity at the head and tail.

inspection period can reduce the peak belt tension by 15% during starting, as the transients that result from break-away are given time to dissipate.

3.2.2.3 Outcomes of Dynamic Analysis

A deeper understanding of the dynamic behaviour in a conveyor system allows the optimisation of several parameters, the most common of which are the selection of the starting and stopping procedures. An example shown below depicts the dynamic behaviour of a loaded 10 km system operating at 4.5 m s^{-1}. The conveyor exhibits a head drive with a 300 second S-type starting procedure, including an inspection period at 0.45 m s^{-1}. This is shown in Figure 3.17, as is the delayed response in the belt at the tail. The movement of the counterweight is shown in Figure 3.18, located adjacent to the drive system at the head. The tension throughout the system during the entire starting procedure is depicted in the surface plot in Figure 3.19. This is perhaps the most useful outcome of dynamic analysis, as peak tensions can be determined and compared with the selected belt. The belt tension at critical locations, such as curves and transitions, can be determined, minimising the risk of excessive belt sag or lift-off.

3.2.3 Belt Conveyor Component Design

In this section the application of simulation methods to the design of critical conveyor components will be presented. The use of the FEM to the design of conveyor drive, tail and snub pulleys, and idler roll seals will be investigated.

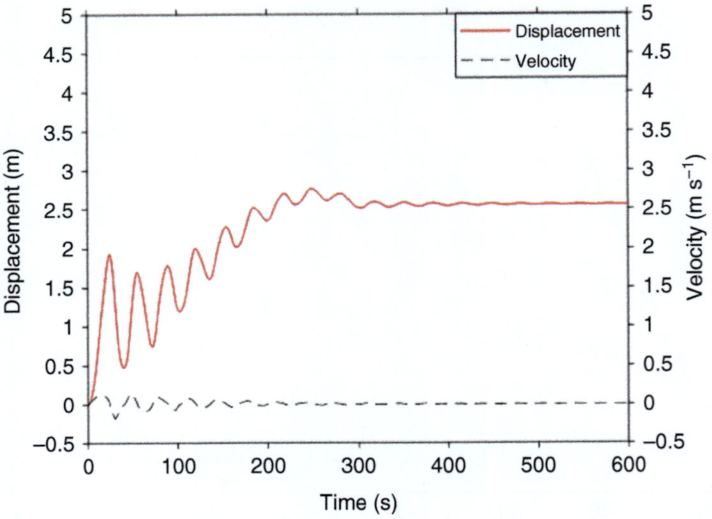

Figure 3.18 Counterweight behaviour.

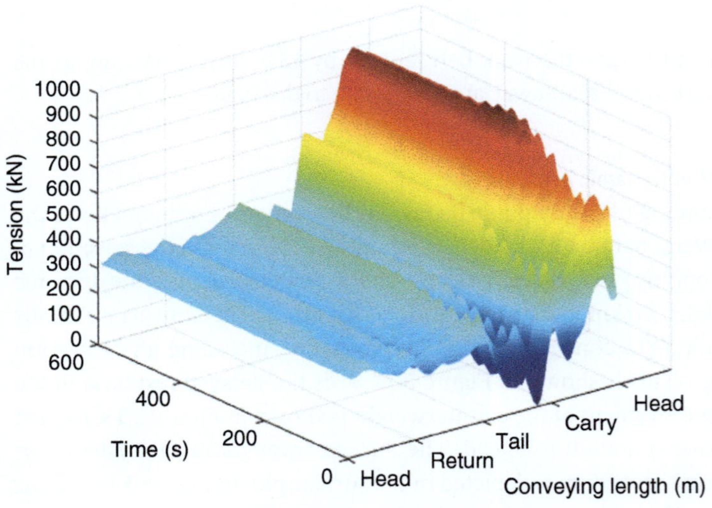

Figure 3.19 Dynamic tensions along the conveyor.

3.2.3.1 Conveyor Belt Pulley Design and Analysis

The main components of an engineered class pulley include the shaft, the drum or shell which is in direct contact with the belt, the end disks which are welded to the ends of the shell, the locking elements to attach the pulley shell to the shaft via hubs welded or machined onto the end disks, the bearing assemblies to support the shaft and lagging attached to the shell surface to increase friction between the shell and conveyor belt. All the components need to function together for reliable, long-term operation. A variety of methods are used by pulley manufacturers and engineering companies to design engineered class pulleys, including purpose-designed software programs that utilise classical mechanics and FEM.

The FEM is a valuable tool to aid in the intricate design of pulleys for belt conveying systems as a detailed stress analysis of the interrelationship between the components can be performed. The FEM can help to optimise the design which can improve reliability and save in cost. As an example, the shape of the end disks can be optimised relatively easily using the FEM to reduce the end disk stiffness, to improve fatigue life, whilst ensuring the optimum distribution of bending stresses throughout the end disk [17].

The following figures show an example pulley that has undergone FEM. For this pulley, the end disk and hub have been machined from a solid plate rather than welded together to minimise the inclusion of stress raisers. The areas of higher stress can be seen in Figures 3.20 and 3.21 for the shaft, Figure 3.22 for the shell, and

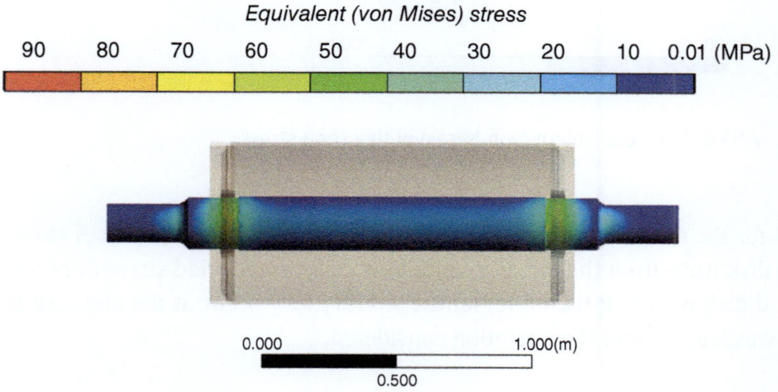

Figure 3.20 ANSYS FEM equivalent (von Mises) pulley shaft stress.

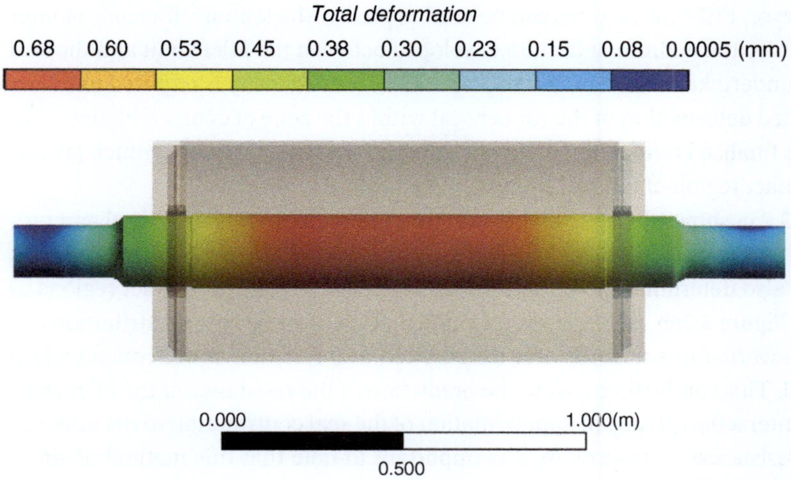

Figure 3.21 ANSYS FEM, total shaft deformation.

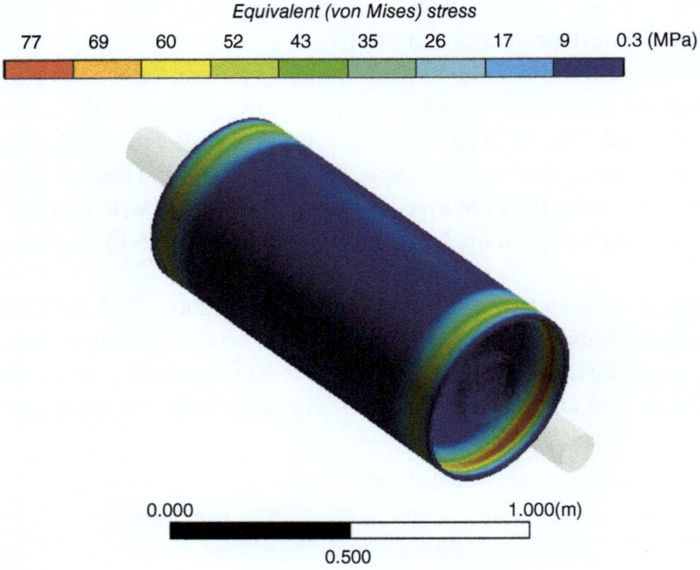

Figure 3.22 ANSYS FEM, equivalent (von Mises) pulley shell stress.

Figure 3.23 for the end disks. The FEM shows that the maximum von Mises stress on the end disk hubs from the locking element is close to the yield strength of the hub and end disk material; therefore, failure would likely occur at the end disk if this pulley was installed for the operation considered.

3.2.3.2 Idler Roll Seal Design and Analysis

The sealing efficiency of idler rolls has a significant influence on the resistance of a belt conveyor. FEM simulations can be used to predict the sealing efficiency of idler seals and also the additional friction developed between the idler shaft and the seal itself. To undertake these simulations, care of the mesh density is required to reflect the expected deformation of the rubber seal within the zone of contact. Figure 3.24a shows the finalised refinement of the mesh where the mesh density is much greater at the contact region of the seal tips (shown in Figure 3.24b).

Figure 3.25a shows the equivalent von Mises stress in the rubber lip seal and provides the best visualisation of the induced stress due to the seal deformation. The FEM can also determine the normal stress distribution over the contact regions as shown in Figure 3.25b. It is important to note that this normal stress distribution can also be converted to a normal force distribution as a function of the contact width of the seal. This can further lead to the prediction of the resistance of the idler shaft and seal interaction giving an approximation of the seal contribution to the idler roll rotating resistance. Furthermore, it is important to note that this method of simulation can be applied to other sealing arrangements in belt conveyor systems where lip seals are used.

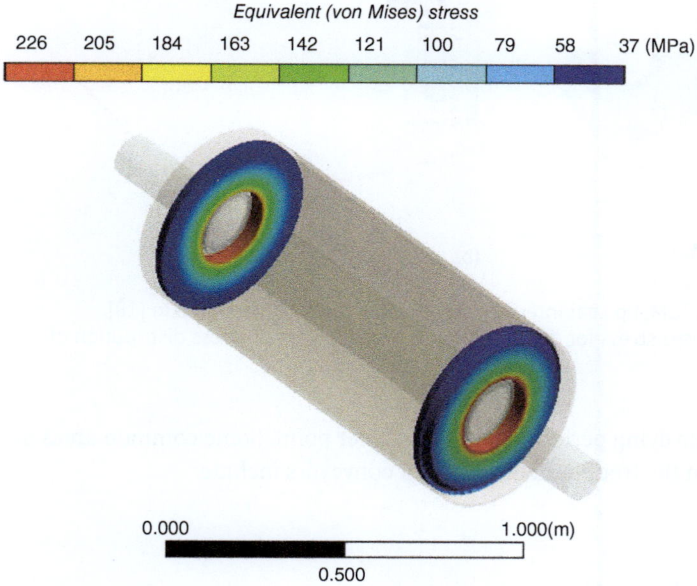

Figure 3.23 ANSYS FEM, equivalent (von Mises) pulley end disk stress.

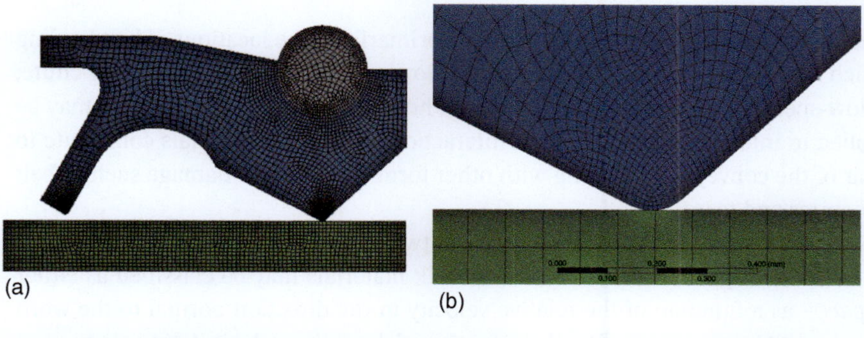

Figure 3.24 ANSYS FEA, lip seal mesh. Source: Donkin [18]. (a) Finalised finite element model mesh. (b) FEA mesh showing detail concentrated at the seal tip.

3.3 Belt Conveyor Troubleshooting

Numerical models may be applied in a troubleshooting capacity to investigate and remedy a range of operating issues in belt conveying applications. The primary simulation technique that is utilised for this purpose is the DEM, focusing on the interactions occurring between bulk material streams and the conveyor belt at a transfer or feeding point. The use of DEM as an analysis tool requires careful consideration of its limitations and a calibration process that is specific to the analysis at hand. If these aspects are considered properly, DEM provides a powerful simulation technique that is suited to the prediction of the complex bulk material flow behaviours

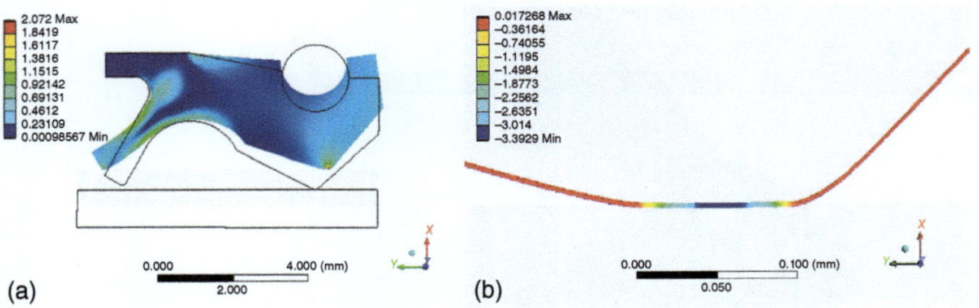

Figure 3.25 ANSYS FEM, lip seal interaction with idler shaft. Source: Donkin [18]. (a) Equivalent von Mises stress for deflection of lip seal. (b) Normal stress distribution of seal tip.

that govern belt conveying performance at a transfer point. Some common areas of DEM application in the troubleshooting of belt conveyors include:

- Belt wear
- Belt tracking
- Spillage of the bulk material and dust issues.

3.3.1 Belt Wear

Wear of the conveyor belt top cover occurs primarily at the locations of belt loading which may involve either high-speed, free-flowing discharge from a transfer chute; or low-speed, high-pressure loading from a hopper onto a belt feeder. DEM may be applied to inform the way in which interactions with bulk materials contribute to wear of the conveyor belt, along with other forms of conveyor damage such as belt punctures and tears.

Wear occurs during interactions between two bodies at a non-zero relative velocity. Wear arising from interactions with bulk materials may be classified as either *Impact* – as a function of the relative velocity in the direction normal to the worn surface or *Abrasion* – as a function of tangential relative velocity, normal pressure and friction and is most often a combination of both mechanisms [19].

By predicting these dynamics, DEM simulations may be applied to inform the design and operation of conveying systems such that belt wear is minimised. This process foremostly considers the design of the transfer chute or hopper that is tasked with feeding bulk material onto the belt, with the aim of reducing the relative velocity between the bulk material stream and the conveyor belt. This process fundamentally depends on the belt velocity that is both the direction of travel, which simplifies to the inclination angle for in-line systems, and the belt speed. The geometry of the transfer chute or feeder interface influences the dynamics of the bulk material at the point of impact with the belt and thereby governs the propensity for wear. The development of a DEM model to predict these characteristics enables an efficient study of the complex interplay between these factors, so that a conveying system can be designed to reduce wear on the belt.

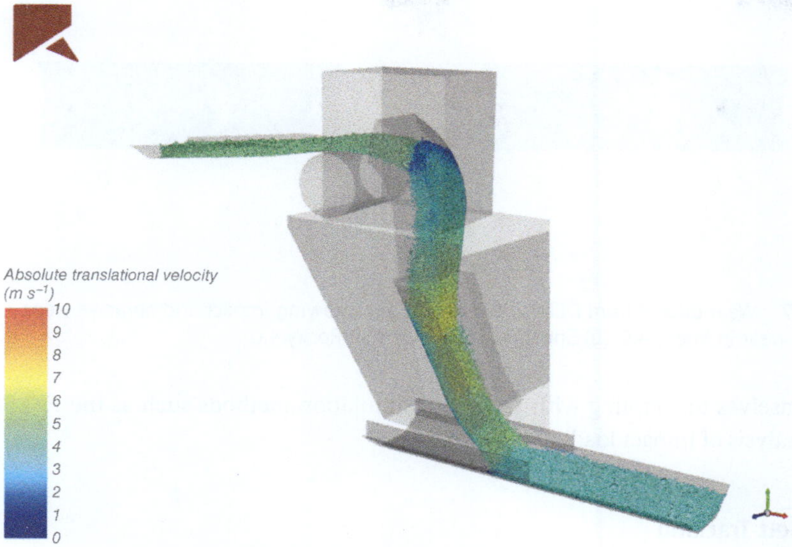

Figure 3.26 Isometric view of Rocky 4.0 DEM simulation showing particle interactions with conveyor belt.

Many commercial DEM packages include analysis features for quantifying the wear conditions arising from the aforementioned dynamics. These typically involve basic models to determine the energy imparted upon the worn surface in both abrasive and impact interactions, as shown in Figure 3.26 [20]. This approach enables the comparison of the wear conditions that arise in different operating conditions, but only in terms of the change in dynamics that influence wear and not by the resultant wear performance of the surface. This approach is useful for relative comparisons between transfer chute or feeder designs and/or operating conditions so that performance may be optimised with respect to belt wear.

Extension of the above comparisons to the prediction of physical resultant wear requires a model that relates the energy imparted by the particles upon the surface, to the surface removal caused by this energy. Many DEM packages have implemented such models, but their application to predict wear in general instances requires caution. Examples of both abrasive (shear) and impact wear predictions within DEM simulations are shown in Figure 3.27. The multifaceted characteristics that influence belt cover wear, including bulk material, belt, environmental and operating properties, are not adequately captured by current wear prediction methods. The development of more rigorous models and subsequent methods for their calibration is an ongoing area of research [21].

While the above commentary has focused on wear, DEM simulations may also be applied to other forms of damage caused by interactions between the conveyed material and the conveyor, such as belt puncture, belt tears or gouging, and impact damage to the idler assemblies. These issues are most prevalent in conveying installations that transport large and irregularly shaped particles, for which DEM approaches are particularly advantageous in predicting flow characteristics. DEM simulations also

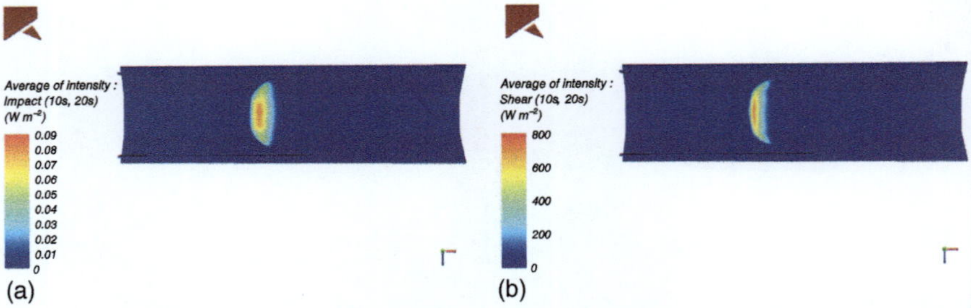

Figure 3.27 Wear output from DEM software package showing impact and abrasive wear. (a) Impact wear in Rocky 4.0. (b) Shear (abrasive) wear in Rocky 4.0.

lend themselves to coupling with structural simulation methods such as the FEM for the analysis of impact loading.

3.3.2 Belt Tracking

Mis-tracking of a conveyor belt occurs when the belt drifts (laterally) from the centreline of the conveyor. This is a common phenomenon and one that causes significant delays in operation and jeopardises the condition of the belt. A range of factors contribute towards belt mis-tracking, with one of the primary causes being non-central distribution of the bulk material across the belt. In this instance the centre of gravity of the conveyed material is offset from the centreline of the belt, causing an uneven lateral reaction from the idler set, resulting in the belt drifting towards the underloaded sided [22].

Non-central presentation of material onto the belt is a function of the transfer chute design and operating conditions. It is most prevalent in transfer chutes that involve a change in conveying direction or a misalignment of the incoming and receiving conveyors. The flow characteristics of such transfers are often inadequately predicted by fundamental analysis methods such as a continuum-based approach, and thereby, benefit from simulation with DEM techniques. The advantages of DEM-based analyses also extend to the simulation of bulk materials that include coarse and irregularly shaped particles. In these instances, DEM may be applied to predict segregation of the coarse and fine fractions, which may impact downstream processing.

The centrality of discharge from a transfer chute is often dependent on the operating conditions. While central loading may be observed for a certain bulk material at a certain throughput, it may not be achieved in alternate conditions. Once a DEM model has been established, the sensitivity of discharge conditions to variations in operating parameters may be readily assessed, as shown in Figure 3.28, often more efficiently than could occur via physical testing methods using scaled conveying geometry. The lateral distribution of material on the receiving conveyor may be assessed quantitatively in DEM by analysis of the mass proportion on each side of the belt and/or by determination of the offset of the conveyed materials centre of gravity to the belt centreline.

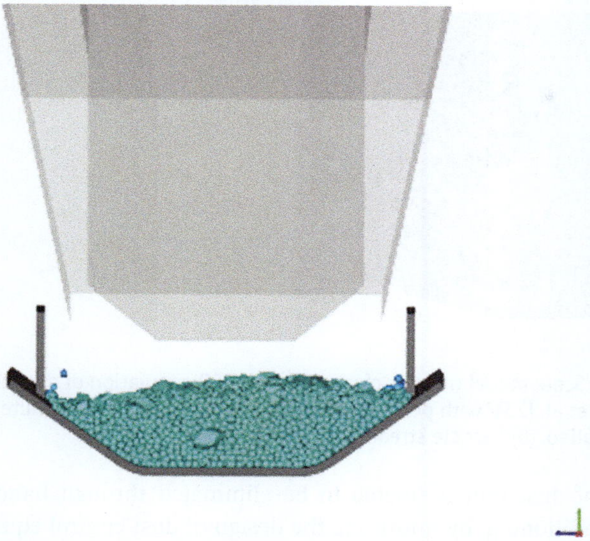

Figure 3.28 Orthographic view of Rocky 4.0 DEM simulation showing non-central discharge conditions.

3.3.3 Spillage and Dust

DEM analysis also lends itself to the prevention of spillage from belt conveying systems. Spillage may occur due to a combination of volumetric and interfacing characteristics, both of which may be informed by DEM analysis. While the volumetric bulk material limits may be readily determined via analytical methods in steady-state operation, the transient conditions that occur during belt loading often benefit from simulation. Further to the conditions described for belt wear in the loading zone, if bulk material is discharged with relative velocity to the conveyor belt, a transition region must exist in which the material is not fully accelerated to the belt speed. To satisfy continuum principles, this results in a greater volumetric requirement in this region than is required in the steady-state conveying region [19]. Simulation of these characteristics, along with particle trajectories in the discharge zone, enables an interface and skirting design that prevents spillage of bulk material. This analysis may also be extended to the prevention of bulk material 'rollback' upon presentation to inclined conveying systems, particularly for coarse particles.

Numerical simulation techniques also find application in the prediction and remediation of dust emissions from conveying systems. DEM and Computational Fluid Dynamics (CFD) methods may be applied in both discrete and/or coupled approaches to analyse dust emissions, with particular application to belt loading locations, as shown in Figure 3.29. The application of these simulation techniques enables optimisation of bulk material interactions through the transfer chute and upon presentation to the receiving conveyor, such that the propensity for dust emission is reduced. These techniques can also be applied to the control and

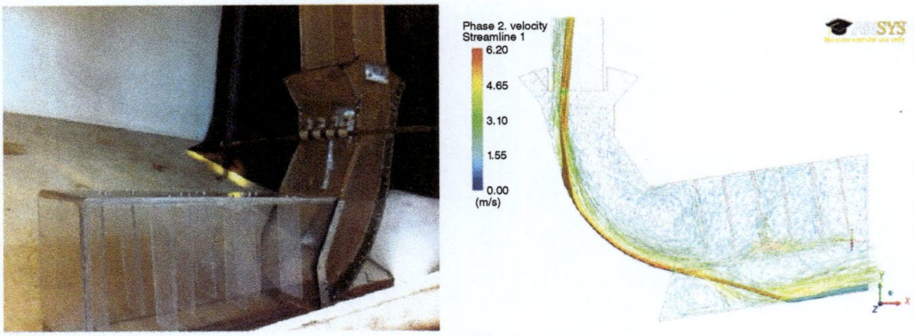

(a) (b)

Figure 3.29 Scale model transfer chute showing CFD simulation of dust emission analysis. Source: Chen et al. [23]/ with permission of, Elsevier. (a) Scale model chute with stilling chamber installed. (b) Particle streamlines and air vectors.

mitigation of dust that is unable to be eliminated through handling design or material conditioning, by informing the design of dust control equipment such as loading hoods and extraction systems.

References

1 Wheeler, C.A. (2016). Rotating resistance of belt conveyor idler rolls. *Journal of Manufacturing Science and Engineering-Transactions of the ASME* 138.

2 Wheeler, C.A. (2003). Analysis of the main resistances of belt conveyors. PhD thesis. University of Newcastle.

3 Munzenberger, P. and Wheeler, C. (2016). Laboratory measurement of indentation rolling resistance of conveyor belts. *Measurement* 94: 909–918.

4 Cook, R.D., Malkus, D.S., Plesha, M.E., and Witt, R.J. (2002). *Concepts and Applications of Finite Element Analysis*, 4e. Wiley.

5 Krause, F. and Hettler, W. (1974). Die belastung der tragrollen von gurtbandförderern mit dreiteiligen tragrollenstationen infolge fördergut unter beachtung des fördervorganges und der schüttguteigenschaften. *Wissenschaftliche Zeitscrift der Technischen Hochschule* 18: 667–674.

6 Spaans, C. (1991). The calculation of the main resistance of belt conveyors. *Bulk Solids Handling* 11 (4): 1–16.

7 Johnson, K.L. (1985). *Contact Mechanics*. Cambridge University Press.

8 Stránsky, J., ed. (2013). Open Source DEM–FEM Coupling.

9 Dratt, M. and Katterfeld, A. (2017). Coupling of FEM and DEM simulations to consider dynamic deformations under particle load. *Granular Matter* 19 (3).

10 ISO 5048:1989. Continuous Mechanical Handling Equipment – Belt Conveyors with Carrying Idlers – Calculation of Operating Power and Tensile Forces. International Organisation for Standardisation; 1989.

11 DIN22101 (2011). Continuous conveyors – belt conveyors for loose bulk materials – basis for calculation and dimensioning. German Institute for Standardisation.

12 Harrison, A. (1984). Dynamic measurement and analysis of steel cord conveyor belts. PhD thesis. University of Newcastle.

13 Lodewijks, G. (1996). Dynamics of belt systems. PhD thesis. Technische Universiteit Delft, The Netherlands.

14 Nordell, L.K., Ciozda, Z. (1984). Transient Belt Stresses during Starting and Stopping: Elastic Response Simulated by Finite Element Methods. Bulk Solids Handling Germany Fed. Repub. 4(1).

15 Robinson, P. (2016). Dynamics of open and closed belt conveyor systems incorporating multiple drives. PhD Thesis. The University of Newcastle, Australia.

16 Nordell LK, editor. (1985) The Theory and Practice of Belt Conveyors Dynamic Analysis, In *Beltcon 4: The International Materials Handling Conference, 7–8 September, 1987.*

17 Sethi V, Nordell LK, editors. (1993) Modern Pulley Design Techniques and Failure Analysis Methods. In *"AIME-SME Annual Convention", in Reno, Nevada of February 1993.*

18 Donkin, P. (2017). Shaft tolerance considerations for the effective performance of rotary lip seals Used in mining applications. MECH4841 Thesis. The University of Newcastle, Australia.

19 Roberts, A.W. (1998). Basic Principles of Bulk Solids, Storage, Flow and Handling. Centre for Bulk Solids and Particulate Technologies, The University of Newcastle, Callaghan, NSW, Australia.

20 Engineering Simulation and Scientific Software (ESSS), Granular Dynamics International, LLC (GDI), and Rocky DEM, Inc. Rocky 3 User Manual; 2015.

21 Reid, S. (2017). Wear of conveyor belt top covers in bulk materials handling applications. MECH4841 thesis. The University of Newcastle, Australia.

22 Conveyor Equipment Manufacturers Association (2014). *Belt Conveyors for Bulk Materials.* Naples, Florida, USA: Conveyor Equipment Manufacturers Association (CEMA).

23 Chen, X.L., Wheeler, C.A., Donohue, T.J. et al. (2012). Evaluation of dust emissions from conveyor transfer chutes using experimental and CFD simulation. *International Journal of Mineral Processing* 110–111: 101–108.

4

Multibody Dynamics and Discrete Element Method Co-Simulations for Large-Scale Industrial Equipment

Dingena L. Schott and Javad Mohajeri

TU Delft, Faculty of Mechanical Engineering, Department of Maritime and Transportation Technology (MTT), Mekelweg 2, NL-2628 CD, Delft, the Netherlands

4.1 Introduction

In the processing and transporting of bulk solid materials such as soil, ores, and powders, equipment and particles interact with each other. Their complex interaction is difficult to capture in analytical models for dynamic interaction problems, and therefore traditionally, the development of handling equipment is done by trial and error of physical prototypes. These design concepts are usually driven by engineering knowledge and in-house experience. It is known that physical prototyping is a time- and cost-intensive design approach that requires multiple concepts to be manufactured and tested on a real scale and under real or realistic operational conditions. Figure 4.1 shows a real-scale grab prototype that was built to be tested at a bulk terminal in 1968.

With the development of particle-based methods such as Discrete Element Method (DEM), it is possible to replace the physical prototype with inexpensive virtual prototypes [1]. This reduces the cost and time required to complete a design iteration, while allowing flexibility and optimisation of designs. Figure 4.2 shows a virtual model of grab that is collecting DEM particles.

For velocity-driven processes, single DEM can be used to simulate the behaviour of a bulk material. By calculating the interaction forces between particles and geometry, the resulting motion of particles can be computed with the help of the equations of motion [3]. After collecting all the information of the particles, it is possible to study the behaviour and flow of a bulk material. This is a computationally intensive method, which has only recently become applicable for large-scale problems with the recent increase in computational power. The forces acting on the geometry can be extracted and used for structural analysis of equipment.

For force-driven processes, the force feedback from the solids on the equipment geometry is required to calculate the motion of the equipment and vice versa repeatedly over time. However, the required single computational method that can capture both particle motions and equipment dynamics does not exist; therefore, separate

Simulations in Bulk Solids Handling: Applications of DEM and other Methods, First Edition.
Edited by Don McGlinchey.

Figure 4.1 Real-scale physical prototype is built to be tested at a bulk terminal (1968, Rotterdam). Source: Courtesy of Nemag.

solvers have to be used instead, i.e. MultiBody Dynamics (MBD), as tool for motion analysis of multibody systems [4]. MBD numerically simulates systems composed of multiple bodies each having mass, inertia, and degrees of freedom [5, 6]. The bodies are connected to each other by means of joints, cables, contacts, or other kinematic or force constraints. The bodies and constraints lead to the equations of motion of the system, which can then be solved.

A combination of both methods could capture both material behaviour and equipment behaviour and predict equipment performance [7]. Figure 4.3 provides an example of applying a DEM-MBD coupling. The MBD module models the equipment components (drive system, wheels, buckets) including their movements, and the DEM module models particles including forces on the bucket geometry.

This work presents the developments in the use of DEM–MBD in particle-based system-related fields and more specifically focuses on the applications in design of handling and processing of bulk. We first show the trends and categorise the domain of particle-based systems. We present a set of reference simulations with analytical solutions that can be used for DEM–MBD coupling verification. Finally, we give an outlook on the role of DEM–MBD in advancing analysis and design.

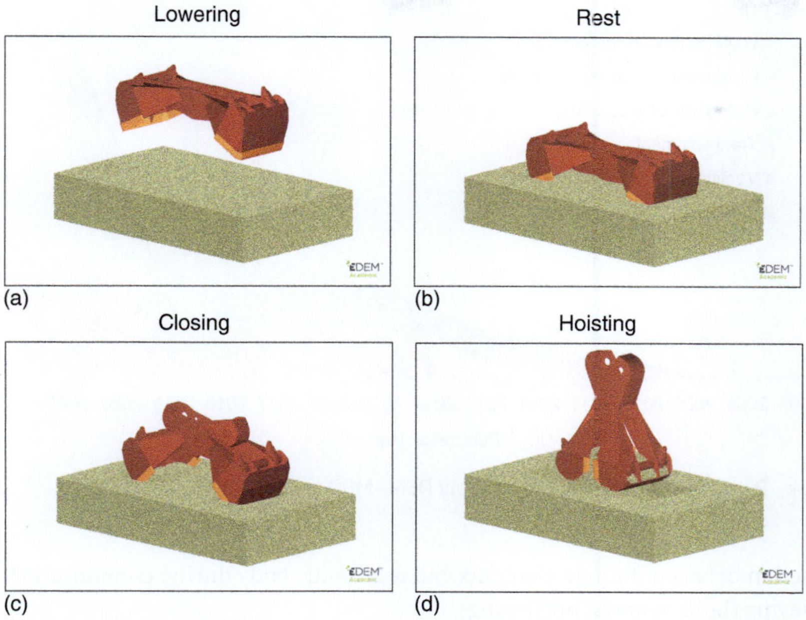

Figure 4.2 Virtual model of the grabbing process by the DEM–MBD co-simulation [1, 2]: (a) lowering, (b) rest, (c) closing, and (d) hoisting.

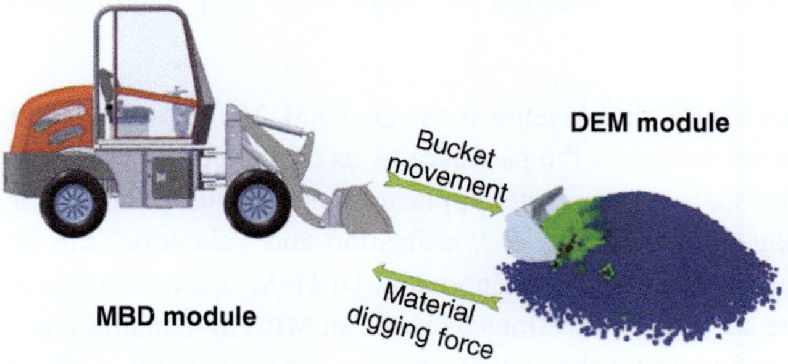

Figure 4.3 Electric loader, an example of DEM–MBD coupling. Source: Li et al. [8]/MDPI/CC BY 4.0.

4.2 Categorisation and Publications on DEM–MBD

To give an overview of the state-of-the-art we performed a literature review on the DEM–MBD resulting in 60 relevant papers using DEM–MBD co-simulation approaches. These papers have been categorised as follows:

- Interaction between a dynamic body and different types of soil with the aim of transportation of the dynamic body by the soil (load-bearing terramechanics).
- Interaction between a dynamic body and different types of soil with the aim of manipulating the soil (load-loosening terramechanics).

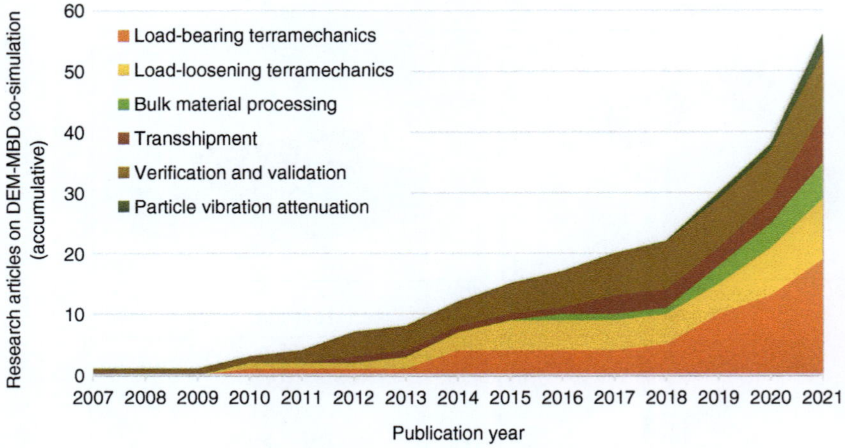

Figure 4.4 Publications that develop or apply DEM–MBD co-simulation.

- Interaction between discrete elements and a dynamic body during comminution and sieving (bulk material processing).
- Interaction between a dynamic body and discrete elements during bulk handling.
- Interaction between damping particles and structures to reduce vibrations (vibration attenuation)
- Development, verification, and validation of DEM–MBD coupling methods (Section 4.3)

The number of research studies that develop or use the DEM–MBD co-simulation has been increased exponentially over the past decade, as presented in Figure 4.4. This figure represents the share of papers per application over time. In the earlier time domain, mainly papers of the category 'Verification and Validation' can be found. This category consists of papers that propose how DEM–MBD co-simulations can be used to model the complex interactions between an MBD system and granular material and remains important throughout the whole time domain. Around half of publications are focused on terramechanics applications, which are also early adopters of the techniques. Bulk material processing and transhipment are also other applications in which a considerable number of research studies has been conducted. An emerging application on using the DEM–MBD co-simulation is the vibration attenuation process, where particles are used to enhance the damping performance of a machine.

In the papers that use or report developments of DEM–MBD co-simulations several commercial and open source software options are used, including EDEM-Adams, EDEM-RecurDyn, EDEM-MotionSolve, Algoryx, LIGGGHTS-Modelica, LIGGGHTS-FMI, and CHRONO. Further details can be found in Tables 4.1–4.5.

In the following text, we will discuss each of the categories in more detail by focusing on the type of studies and category of the parameters investigated. Because the verification and validation of DEM simulations is often not or not rightly included

Table 4.1 Detailed overview of literature on load-bearing terramechanics using DEM–MBD.

				DEM			MBD		
Author	**Application**	**Scale**	**Validated KPIs**	**Solver**	**Bulk solid**	**Particle shape**	**Calibrated**	**Solver**	**Motion input**
Edwards [9]	Wheeled and tracked vehicles	Full *in situ*	None	EDEM	Loose sand	Sphere	No	MSC ADAMS	(–)
Pesheck et al. [10]	Off-road vehicle mobility	Full *in situ*	Vehicle acceleration	EDEM	Fine sand	Sphere	Yes	MSC ADAMS	Steering control
Shin et al. [11]	Single wheel	Lab	Sinkage	Xdynamics	Sand	Cluster	No	Xdynamics	Wheel speed
Teslanko and Belyakov [12]	Motor transport vehicle	1 : 43	(–)	EDEM	Snow	Sphere	(–)	MSC ADAMS	Virtual engine
Chen et al. [13]	Electric shovel	Full	None	EDEM	Sand-gravel	Cluster	No	RecurDyn	Virtual engine
Nakashima et al. [14]	Single wheel (lunar rover)	Lab	slope angle, travel distance, torque	Self-developed	Lunar regolith simulant	Sphere	Yes	Self-developed	Force
Toesen et al. [15, 16]	Helical design	Lab	Craft speed, power, velocity	EDEM	Glass beads	Cluster	No	MSC ADAMS	Rotational speed
Ji and Liang [17]	Lunar lander	Full	leg force, immersion depth	EDEM	lunar soil	Sphere	No	(–)	Gravity force
Liu et al. [18]	Ballasted railway track	Full	Sleeper acceleration	(–)	Railway ballast	Cluster	(–)	(–)	Moving load
Dai et al. [19]	Deep sea mining	Lab	(–)	Self-developed	Riser	Rigid-flexible coupled	(–)	RecurDyn	(–)
Dai et al. [20]	Deep sea mining	Lab	Sinkage pressure	Self-developed	Seabed sediment	Cluster	Yes	RecurDyn	Speed controller
Thoeni et al. [21]	Waste rock barrier	Full	Weight distribution of truck	Algoryx simulation	Rock	Sphere	Yes	Algoryx simulation	Velocity

(–) missing information.

Table 4.2 Detailed overview of literature on load-loosening terramechanics using DEM–MBD.

				DEM			MBD		
Author	**Application**	**Scale**	**Validated KPIs**	**Solver**	**Bulk solid**	**Particle shape**	**Calibrated**	**Solver**	**Motion input**
Ahmadi Ghoohaki [22]	Backhoe digging	Full	None	DIS	Sand	Cluster	Yes	DIS	Linear and rotary
Gajanan Sane [23]	Bulldozer	Full	None	DIS	Cohesive soil	Cluster	Yes	DIS	Linear and rotary
AXAS [24]	Roller compaction	Full *in situ*	Contact stress	Demify	Gravel	Cluster	Yes	Self-developed	Virtual engine
Shi et al. [25]	Ballast bed	Full *in situ*	Lateral resistance	EDEM	Ballast	Cluster	No	MSC ADAMS	Tamper velocity
Shi et al. [26]	Ballad bed	Full	None	EDEM	Ballast	Cluster	Yes	MSC ADAMS	Tamper velocity
Lichtenheldt et al. [27]	HP-Mole	Lab	Deflection	(–)	Mars soil	Sphere	No	(–)	DC motor
Alkalla et al. [28]	DROD	Lab	Only the MBD model	EDEM	Regolith	Cluster	No	MSC ADAMS	DG geared motor

(–) missing information.

Table 4.3 Detailed overview of literature on bulk material processing using DEM–MBD.

				DEM			MBD		
Author	**Application**	**Scale**	**Validated KPIs**	**Solver**	**Bulk solid**	**Particle shape**	**Calibrated**	**Solver**	**Motion input**
Cheng et al. [29, 30]	Cone crusher	Full *in situ*	Displacement, power, PSD	EDEM	Marble	Cluster	Yes	RecurDyn	Drive speed
Barrios and Tavares [31]	Roll grinding	Full *in situ*	(–)	EDEM	Iron ore	Sphere	Yes	Self-developed	Roll velocity
Rodriguez et al. [32]	Roll grinding	Full *in situ*	PSD	EDEM	Pellet feed	Sphere	Yes	Self-developed	Roll velocity
Wang et al. [33]	Screening	Full *in situ*	PSD	AgriDEM	Soybean	Sphere	(–)	(–)	Velocity
Wu et al. [34]	Screening	Full	None	EDEM	Coal	Sphere	Yes	RecurDyn	Velocity

(–) missing information.

Table 4.4 Detailed overview of literature on bulk material handling using DEM–MBD.

				DEM			MBD		
Author	**Application**	**Scale**	**Validated KPIs**	**Solver**	**Bulk solid**	**Particle shape**	**Calibrated**	**Solver**	**Motion input**
Ahmadi Goohaki [22]	Backhoe digging	Full	None	DIS	Sand	Cluster	Yes	DIS	Linear and rotary
Gajanan Sane [23]	Bulldozer	Full	None	DIS	Cohesive soil	Cluster	Yes	DIS	Linear and rotary
Hess et al. [35]	Mobile excavator	Full-scale	Dynamics, kinematics,	LIGGGHTS	Soil	Sphere	Yes	Modelica	Control and hydraulic system
Schott et al. [36]	Grabs	Full *in situ*	Dynamics, kinematics	EDEM	iron pellets	Sphere	Yes	MSC ADAMS	Crane winch
Mohajeri et al. [37]	Grabs	Full *in situ*	Dynamics, kinematics, density, volume	EDEM	Cohesive iron ore	Sphere	Yes	MSC ADAMS	Crane winch
Mohajeri et al. [38]	Grabs	Full	None	EDEM	Cohesive iron ore	Sphere	No	MSC ADAMS	Crane winch
Fimbinger and Becker [39]	Hay Baler belt system	Full	None	ThreeParticle/ CAE	Various Granular, hay	Various	No	ThreeParticle/ CAE	Pully
Li et al. [8]	Electric loader	Full	None	EDEM	Coal, iron ore, lime stone	Cluster	Yes	RecurDyn	Engine and hydraulic system

Table 4.5 Detailed overview of literature on vibration attenuation by particle systems using DEM–MBD.

				DEM			MBD		
Author	**Application**	**Scale**	**Validated KPIs**	**Solver**	**Bulk solid**	**Particle shape**	**Calibrated**	**Solver**	**Motion input**
Chung and Wu [47]	Gear transmission	Full	None	NA	Steel beads	Sphere	No	NA	Shaft rotational velocity
Wu et al. [48]	Particle-based damper	Lab scale	Position	PFC3D	Steel beads	Sphere	No	Self-developed	Vibration
Dan et al. [40]	Cable vibration	NA	Vibration Frequency	EDEM	Damper	Sphere	No	MSC ADAMS	Cable vibration

in DEM-related publications, we present ways of developing a coupling including verification approaches in more detail in Section 4.3.

4.2.1 Load-Bearing Terramechanics

The largest field of application is load-bearing terramechanics. Terramechanics is the study of soil in interaction with equipment. Load-bearing terramechanics particularly considers the interaction of wheeled or tracked vehicles on various surfaces; these are relevant for domains such as agriculture, construction, space, and sea beds. The soil can be modelled as a range of small particles and the vehicle as a set of dynamic bodies. DEM–MBD is required to determine the forces acting on the interface, the large displacement of the simulated soil, and the forces acting on the bodies.

Historically, terramechanics have been defined by semi-empirical equations to predict various mobility aspects. These empirical equations were able to derive wheel torque, thrust, and sinkage by estimating stress distributions along the wheel-terrain contact region. Kobayashi et al. [41] summarised classical terramechanics calculation strategies and described the limitations of these models of Bekker and Wong-Reece. The introduction of DEM–MBD co-simulation offers a more accurate alternative to these traditional equations; it enables three-dimensional studies with adequate pressure-sinkage rates for slipping wheels and allows to evaluate slip-induced mobility chances [10, 41, 42]. Table 4.1 provides a detailed overview of research studies that used DEM–MBD co-simulation for studying load-bearing terramechanic applications.

Vehicle applications, i.e. tracked and wheeled vehicles on a range of surface materials are studied. The steady-state single wheel test interaction on Mojave Martian Simulant was modelled and compared with experiments and Bekker-type methods [42]. This model is used for analyses of the interaction between a wheel and sand particles [11], interaction between a vehicle and snow [12], behaviour of tracked and wheeled vehicles over soft soil terrains [9], design and mobility of helically driven vehicles in granular materials by varying screw geometries and rotational speeds [11] (as shown in Figure 4.5), to predict the capability of vehicles to traverse on terrains consisting of various soil types [10], design optimisation of tracked chassis

(a) (b)

Figure 4.5 (a) Single-Wheel test bed [42], and (b) Simulation of helically driven vehicles. Source: Smith et al. [15]/with permission of Royal Society of Chemistry.

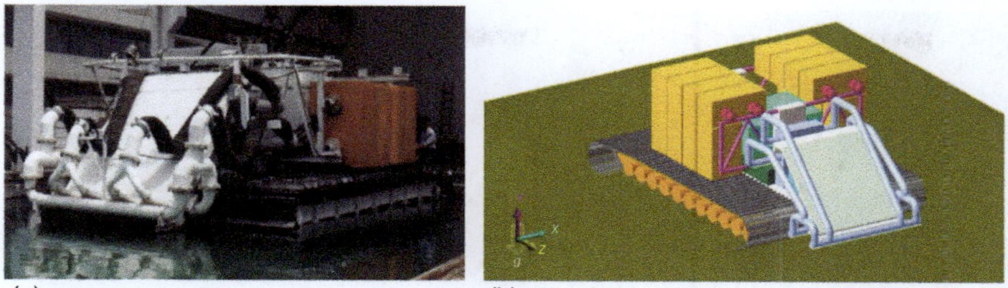

Figure 4.6 Seabed miner (a) and its MBD model (b). Source: Dai et al. [19]/IEEE/CC BY 4.0.

performance of an electric shovel [43], and design of waste rock barriers that serve the purpose of a safety berm which can prevent haul trucks from fatal accidents [21].

Applications in space with different gravitational conditions can be easily altered in the simulation environment included for example analysis of the single wheel performance of a lugged wheel for a small lunar rover on a sloped terrain [14] and investigation of the landing process of a lunar lander to guarantee safe landing using DEM–MBD–FEM [17].

For railway applications the impact of the vibration energy due to repeated vehicle loads of a high-speed railway on a ballasted track structure was investigated [18]. By in-field experiments and with the help of DEM–MBD co-simulation, the transfer behaviour of vibrations and the main vibration frequencies of the ballast at the concrete sleepers were discovered. By examining these frequencies, the main cause for a change in the structural health state of the ballasted track could be found

For offshore applications the coordinated motion control of a seabed miner by considering its interaction with the seabed sediment and the fluid dynamics at the deep-sea level has been explored [19, 20]. Using computational fluid dynamics (CFD) for the fluid phase to determine hydrodynamic loads on the miner, DEM for the seabed sediment and an MBD model of the deep-sea mining vehicle has been established [15]. Figure 4.6 shows a seabed miner and its DEM model. Having a DEM–MBD co-simulation allows introduction of an adaptive PID controller to improve the motion accuracy of the seabed miner, thereby providing a theoretical basis for prediction and optimisation of the performance of a seabed miner [19].

4.2.2 Load-Loosening Terramechanics

Another important application field is that of terrain manipulation (load-loosening terramechanics). In this category, the interaction between terrain manipulation equipment (e.g. ploughs and bulldozers), and the terrain is studied. Compared to the load-bearing terramechanics, the relocation of the land will be much greater. A common example of soil manipulation is soil-tillage [44]. Tilling is the act of turning over and breaking up the soil and is needed to mix amendments into the soil. This involves interaction between a dynamic body and a large amount of soil particles and can therefore be modelled using a DEM–MBD co-simulation. It requires to consider variability of soil media, determination of micro-mechanical

(a) (b)

Figure 4.7 (a) Tamping operation, and (b) Tamping co-simulation. Source: Shi et al. [25]/with permission of Elsevier.

model parameters and the modelling the process of breakage and formation of agglomerates [45]. Table 4.2 provides a detailed overview of research studies that used DEM–MBD co-simulation for studying load-loosening terramechanic applications.

The digging operation of a backhoe was modelled to improve the performance of heavy duty earth-moving equipment [22]. The backhoe has been modelled as different rigid bodies (e.g. frame, manipulators links, track segments, wheels, and sprockets) connected to each other using joints. Sand has been used as a digging material, and its material model with cubic-shaped particles was validated using a conical hopper sand flow experiment. By co-simulation of sand and backhoe model vehicle stability, the required actuator and joint forces while digging could be determined. The excavation and moving operation of cohesive soft soil (e.g. mud and snow) by a bulldozer was modelled using an explicit MBD finite element code with integrated DEM [23]. The model can be utilised to predict motion, loads, and required actuator forces that can be used to improve the design of bulldozer components.

Next to ploughing and excavating, co-simulation is used for compaction of soil and roads using road compactors [24]. Well-compacted roads are more resistant to weather influences and traffic loads and will therefore last longer. Also, as shown in Figure 4.7, models for the creation of (railway) ballast beds using tamping operations to improve the mechanical quality of the beds allowed the optimisation of tamping parameters [25] and improvement of the service lifetime of the ballast bed [26].

By using DEM–MBD co-simulations, the interaction between a soil and a drill can also be modelled and can be applied for different conditions (e.g. on other planets) [27, 28].

4.2.3 Bulk Material Processing

This category focuses on the interaction between a dynamic body and granular material during value-added activities such as communition i.e. particle size reduction by crushing and grinding, and screening of bulk materials to separate particulate materials according to their size. Table 4.3 provides a detailed overview of research studies that used DEM–MBD co-simulation for studying bulk material processing applications.

The dynamic response of inertia cone crushers was modelled by describing mechanical motions, non-linear contacts and loading responses as a function of various geometric, operational, and material parameters [29, 30]. The industrial-scale experiments indicated that the displacement and throughput capacity are in good agreement with the simulation output and that the prediction in terms of power draw and critical compression ratio should be improved [25]. Investigations of the influence of the fixed cone mass and the moving cone mass resulted in the introduction of a new dynamic balancing mechanism resulting in a fixed mass reduction [26], improving the product size distribution, and decreasing cone masses; as a result, a signification reduction in manufacturing costs was achieved. Specific recommendations include decreasing running costs for users and focusing on operational variables such as the optimum drive shaft speed, the eccentric distance of the eccentric vibrator, and the discharge gap.

Research on high-pressure grinding rolls (HPGRs) using DEM–MBD focused on design variables of the hydraulic system (e.g. accumulator pressures and volumes) together with system parameters (e.g. roller geometry, roller design, and material loading response) have been used as an input to determine operating outputs (e.g. material throughput, operating gap, energy consumption, and roller pressure distribution) [31]. Investigation on the effect of grinding roll geometry on performance and on roll wear revealed that the conducted simulations predicted a less severe reduction in fineness of the product with roll wear than has been observed in practise [32]. Figure 4.8 shows an example of applying DEM–MBD co-simulation for studying HPGR application.

The effects of structure parameters of a swing-bar sieve (e.g. inclination angle, crank length, and vibration direction angle) and the effect of operating parameters (e.g. crank speed and feed rate) on the screening deck performance were studied by comparing the passing percentage of different sizes of soybeans in the simulations to experimental results [33]. The results were used to develop an instructive way for designing and optimising a swing-bar sieve.

A two-way particle-screen panel coupling strategy has been proposed to consider the effect of particles on flip-flow screen panels together with the effect of flip-flow

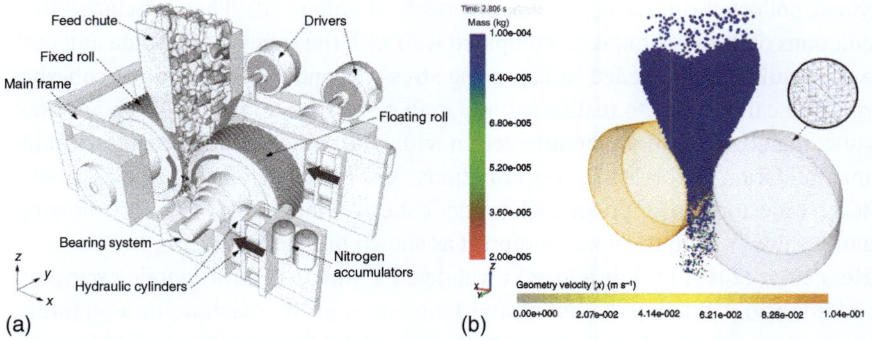

Figure 4.8 High-pressure grinding rolls (HPGR) schematics (left) and model (right). Source: Barrios and Tavares [31]/With permission of Elsevier.

screen panels on particles [34]. Investigations included the screening efficiency, the production ratio, and the vibrating velocity of the particles and screen as a function of the rotational engine speed. The insights gained are helpful in the research and development of screening theory and flip-flow screens.

4.2.4 Bulk Material Handling

This category focuses on the interaction between particles and dynamic bodies during the handling and transportation of bulk material from one means of transport to the other. The work categorised here is exploring ways to increase the capacity and efficiency of bulk material handling equipment (e.g. conveyors, stacker reclaimers, grab unloaders) while keeping the operating costs as low as possible. Table 4.4 provides a detailed overview of research studies that used DEM–MBD co-simulation for studying bulk material processing applications.

A scissors grab model has been developed through a coupling between DEM and MBD (defined by [44]) and was validated by industrial-scale experiments [46]. The model has been validated by comparing the predicted torques by the model on the winches of the grab with measurements. With the validated model, a better understanding of factors influencing grab performance was gained, and this knowledge was used in the design of new-generation grabs [36]. Here, kinematics tracking techniques and comparison of the measured experimental and model data were used to confirm adequate model performance. The use of these models in a virtual environment as a virtual prototyping tool avoids expensive, time-consuming and resource wasting trial and error approaches.

Further steps in grab design were taken by including cohesive and stress history-dependent materials [37]. Here, a validated DEM–MBD co-simulation was developed that included the material behaviour by using an elastoplastic adhesive contact model representing cohesive and stress history-dependent iron ore. This way of modelling includes the effect of pre-consolidation during loading and sailing, where equipment performance at the receiving ports depends on the consolidation level. The validation was established by in-field full-scale measurements and recordings (e.g. force in cables, torque in winches, kinematics of geometry, payload, collected volume and average porosity of bulk solid). They concluded that predictions of the co-simulation compared well with the experimental data and that the co-simulation succeeded in capturing stress-dependent behaviour of cohesive cargo, and can be used to realise optimal grab performance. Figure 4.9 shows how distinct material behaviours in interaction with grab could be captured successfully using validation DEM–MBD co-simulations; the footprint of grab is close to its cutting trajectory in a pre-consolidation cohesive material, while free-flowing materials flow within the grab's footprint as shown in Figure 4.9b.

Hess et al. (2016) [35] developed a validated simulation for a mobile excavator. In addition to validating dynamics and kinematics of the machine in a quantitative way, two different filling situations are compared between the simulation and the experiment, as shown in Figure 4.10. The approach to treat the DEM and MBD

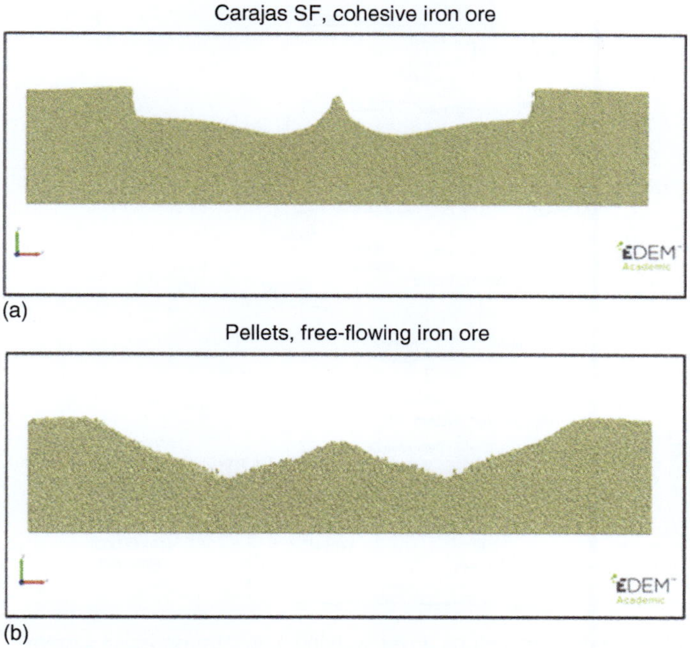

Figure 4.9 Footprint of the grabbing process on two types of iron ore: (a) Carajas SF and (b) Pellets. Source: Mohajeri et al. [37]/With permission of Elsevier.

Figure 4.10 Comparing filling situations of the mobile excavator in experiment and simulation. Source: TU Dresden, Chair of Construction Machinery in Hess et al. [35].

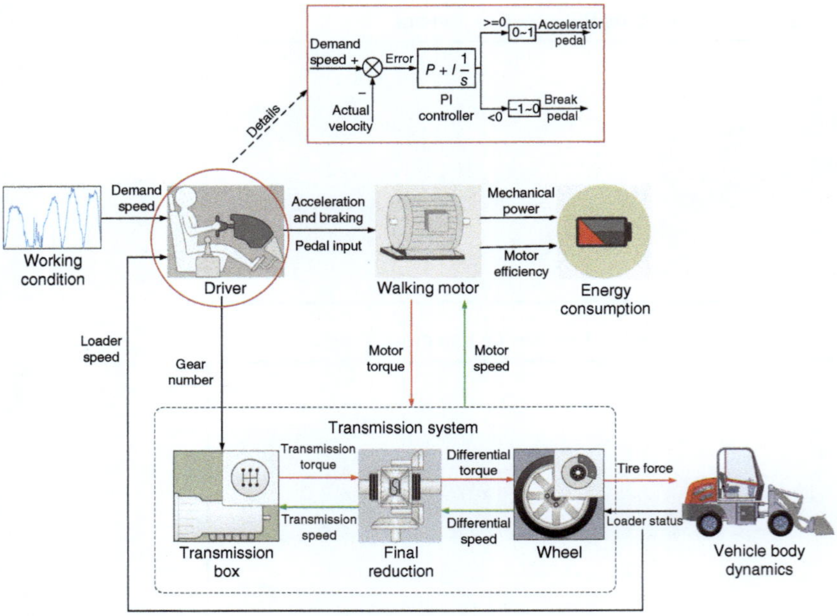

Figure 4.11 Schematic diagram of modelling driver, walking motor, transmission system for an electric loader. Source: Li et al. [8]/MDPI/CC BY 4.0.

sub-models individually and calibrate them in their own field of application is found to be a proper approach. Such a co-simulation can be used to optimise the machine design and operation.

Another application is assessing the operating performance of pure electric loaders used in construction work as an effective energy-saving and emission-reduction alternative to conventional loaders [8]. They used DEM–MBD co-simulation to compare the performance of different types of electrical motors. Figure 4.11 shows how the driver, walking motor, and transmission system are modelled within a DEM–MBD co-simulation. Fimbinger and Becker [39] used DEM to model a hay bale with elongated particles, a bonded particle model for the belt, and the MBD for idlers and the driven pulley.

4.2.5 Vibration Attenuation

Although not directly related to the manipulation of large quantities of granular materials such as previous categories, the field of vibration attenuation also explores the use of granular media to reduce vibration in gear transmission [47, 48] and cable dampers when external influences such as weather conditions cause excitation [40]. Such particle systems can attenuate vibrations due to their damping properties. Table 4.5 provides an overview of research studies that used DEM-MBD co-simulation for studying vibration attenuation.

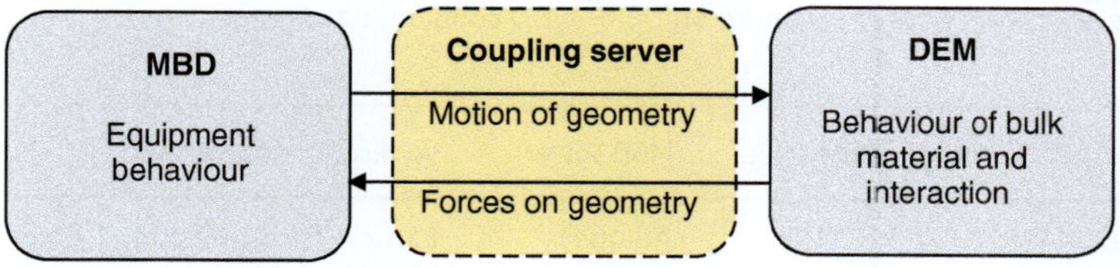

Figure 4.12 Schematic of the coupling server sharing information between using MBD and DEM.

4.3 Developing a DEM and MBD Coupling

The main purpose of linking the bulk material simulation to the equipment simulation is to perform the equipment simulation with loads caused by the bulk material. A coupling is essential for two reasons. First, the discrete element software assumes the contact areas between particles and geometry as rigid bodies without a mass. Therefore, dynamic inertia properties of the dynamic body cannot be described. Second, the MDB software is unable to describe particle interactions since it cannot detect particle-to-particle and particle-to-geometry contacts. The DEM solver is able to solve the equations of motion of simple, discrete bodies in direct contact, but it is not suitable for bodies that are in mutual contact due to constraints (e.g. rods, joints, and force elements). For this reason, there has been a great interest for coupling the two separate solvers in a systematic way (Figure 4.12). Developing the coupling requires verification to assure that ideas are accurately and faithfully translated into computer code or mathematical formalisms or in other words "a demonstration that the modelling formalism is correct." [49].

Lommen et al. [44] documented the development of their coupling server in detail, and provided test cases that can be used for verification of any software combination. Starting point for the coupling is a vector with the position and trajectory $x(t)$ of a coupled body during a communication interval nDt, where Dt is the DEM time-step. This vector contains the position, orientation, and velocities of the geometry. With this data of the geometry, the DEM simulation checks whether particles are in contact with the body during the communication interval and calculates the corresponding load data.

The load data of each coupled body consist of six components: both forces and torques in three directions. The forces and moments are oriented in the directions of the global coordinate system and around the body's centre of mass. At each communication interval, the load data are requested from the DEM simulation for all the bodies to be coupled.

The MBD solver calculates the movement of the mechanism based on all loads on all external loads on the mechanisms including those from DEM. This process is repeated until the simulation time has been reached. A collection of papers developing the coupling with various software combinations is included in Table 4.6.

Table 4.6 Detailed overview of literature on DEM–MBD coupling development and verifications.

Application	Year	DEM solver	MBD solver	Verification test case
Lommen et al. [44]	2018	EDEM	Adams	Particle-cube collision; Motorised rotating pendulum, Spring-damper system
Elmqvist and Ghandriz [50]	2015	Modelica	Modelica	Multiple working examples, but no verification
Chen et al. [51]	2017	LIGGGHTS	Modelica	Sand agitation with a pneumatic motor
Richter et al. [52]	2020	LIGGGHTS	FMI	Pendulum impact; rotatory cross and bucket test rig
Yoon et al. [53]	2011	SAMADII	RecurDyn	Bulk weight
Ortiz et al. [54]	2016	EDEM	MSC ADAMS	Vehicle on granular terrain, not verified
Sanborn et al. [55]	2021	Own code	RecurDyn	Weight balance, car collision, pendulum, compressor

4.3.1 Verification Tests

For verification, four different tests are proposed, starting with a simple test and gradually increasing to the complex. At each simulation experiment, the simulation results are compared to the analytical solution. Without such a verification, it will remain unclear whether the implemented coupling works as intended. The four tests have been selected to test the implemented coupling step by step, with each test focusing on additional aspects:

- The first test examines whether the implemented coupling can accurately predict the contact of a single particle colliding with a wall of a coupled body. Also, this test examines if the coupling server is interfacing correctly with MBD code. A known load of particle collisions is applied to the body and MBD code has to return the corresponding behaviour of the body. This test is limited to translation only.
- The second test is to verify that the coupling server interfaces correctly with the DEM code. The DEM code will be given a known motion for a body and has to return the corresponding load data to the coupling server. The given motion consists of both translations and rotations.
- The third test verifies that the two-way coupling works for a scenario limited to translational forces and movements.
- The final test of the coupling of DEM and MBD method verifies that the two-way coupling works for a scenario that includes rotations and moments.

The agreement between the co-simulation results and the analytical results is suggested to evaluate using the coefficient of determination R^2 as described by Weisberg [56]. Because both the results are based on the same mathematical

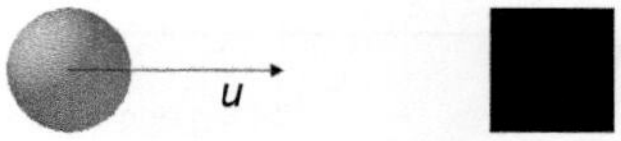

Figure 4.13 Particle cube collision test. Source: Lommen et al. [44]/Emerald Publishing Limited/CC BY 4.0.

problem, a very high coefficient of determination is to be expected. R^2 should exceed 0.99, demanding high accuracy while allowing for numerical scatter in the co-simulation.

4.3.1.1 Particle–Wall Collision

The first test of the verification process is a simple test where particles are shot at and collide with a geometry. Instead of using two particles in DEM, one of them has been replaced with a body whose behaviour is calculated with MBD. The aim of this test is to investigate whether a simple collision between particle and geometry can be computed correctly. In this first scenario, displayed in Figure 4.13, a single particle was generated and collided with the cube, which has only translational degrees of freedom. During the simulation, where gravitational forces are absent, the particles collided with the cube, transferring their energy according to Eqs. (4.1) and (4.2):

$$v_c = \frac{m_p u_p + m_c u_c + m_p C_R (u_p - u_c)}{m_p + m_c} \tag{4.1}$$

$$v_p = \frac{m_p u_p + m_c u_c + m_c C_R (u_c - u_p)}{m_p + m_c} \tag{4.2}$$

where v_p is the velocity after collision, u_p the initial velocity, and m_p the mass of the particle and v_c, u_c, and m_c for the cube.

Next, the single particle is replaced with a stream of particles hitting the cube with their own initial speed minus the current speed of the cube. Because of the incoming momentum of the particles, the cube starts accelerating, although the rate of acceleration decreases because of the increasing distance between the starting point of the particles and the position of the cube, thereby requiring the particles to travel longer before they collide with the cube. Figure 4.14 shows the velocity and displacement of the cube calculated by the co-simulation and theoretical calculations based on Eq. (4.1). Lommen et al. [44] showed that the agreement between co-simulation and theory is very good as the coefficients of determination R^2 exceed 0.999, thus proving that the collision between the particle and the body is computed correctly.

4.3.1.2 Pendulum

The second test of the verification process used a box filled with particles connected to a pendulum, rotating at a constant angular velocity (Figure 4.15a). The aim of this test is to assure that the coupling server correctly updates the position of the body in DEM and whether the load calculated in DEM corresponds to the theoretical approach. This is achieved by comparing the forces acting on the joint of the pendulum predicted by the simulation to the gravitational and centrifugal forces of the box with particles. As the interaction forces do not affect the motion of pendulum, the forces on the joint and the torque required for the rotation can be calculated

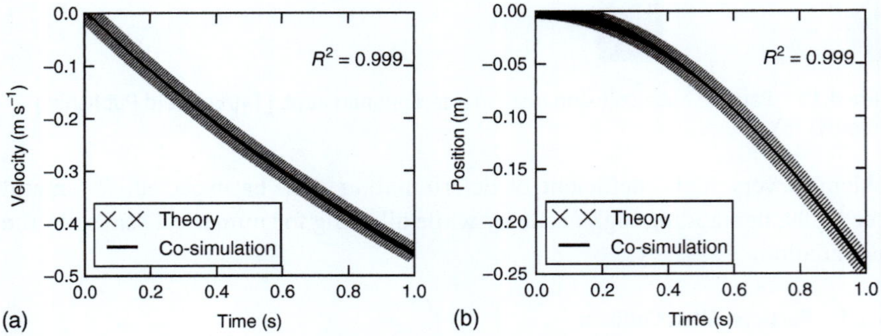

Figure 4.14 Comparison of theory and co-simulation results of velocity and displacement of the cube hit by a stream of particles (particle mass of 0.01 kg) with an initial speed of 2 m s^{-1} minus the current speed of the cube. Particles are created at the rate of 200 particles per minute. Source: Lommen et al. [44]/Emerald Publishing Limited/CC BY 4.0: (a) particle velocity and (b) particle position.

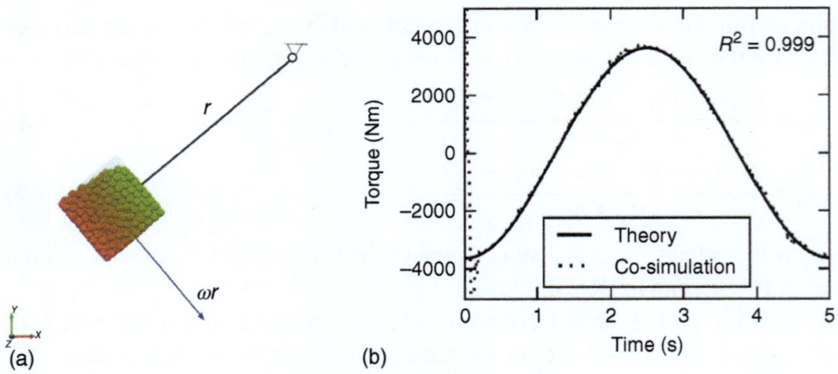

Figure 4.15 Motorised pendulum with mass *m* according to Eq. (4.6), consisting of m_p of 236.6 kg and m_c of 10 kg. The pendulum has a horizontal starting position and rotates counter-clockwise at a constant angular velocity *v* of 72 deg s^{-1}. The radius *r* of the pendulum is 1.5 m, and gravitational constant equals 9.81 m s^{-2}. Source: Lommen et al. [44]/Emerald Publishing Limited/CC BY 4.0: (a) box with particles moving in a circular motion, and (b) torque required for the circular motion.

by Eqs. (4.3)–(4.5):

$$F_x = -m\omega^2 r \cos\theta \tag{4.3}$$

$$F_y = m\omega^2 r \sin\theta + mg \tag{4.4}$$

$$T_z = F_x r \sin\theta - F_y r \cos\theta \tag{4.5}$$

$$m = \sum m_{\mathrm{p}} + m_{\mathrm{c}} \tag{4.6}$$

Figure 4.15b shows the torque required for rotating the particle box at the prescribed angular velocity. After 0.25 seconds, the box in the simulation was

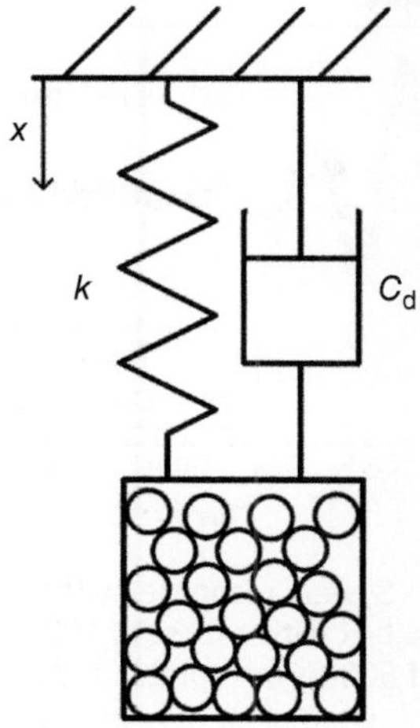

Figure 4.16 Schematic of translational spring damper system. Source: Lommen et al. [44]/Emerald Publishing Limited/CC BY 4.0.

accelerated to the prescribed constant angular velocity. The calculations by the co-simulation match very well with the theoretical approach based on Eq. (4.5) (with an R^2 of 0.999) for a given input for both translational movements and rotational movements.

4.3.1.3 Translating Spring Damper System

The third test of the verification process uses a box of particles connected to a translational spring damper system (Figure 4.16) limited to translational forces and movements. The aim of this test is to investigate whether the coupling server and MBD process the load input correctly and if it results in the known system response as described by Eqs. (4.6) and (4.7):

$$m\frac{d^2x}{dt^2} + C_d\frac{dx}{dt} + kx = mg \tag{4.7}$$

Figure 4.17 shows the velocity and position of the box of particles during the simulation. Because of weight of the particles and the box, the box drops down in search for a new equilibrium. Because of the damper present in the system, the velocity is reduced, damping the system and gradually slowing down the box. It can also be observed from Figure 4.17 that the response predicted by the co-simulation correlates well with the response following from Eq. (4.7). This proves that the coupling server and MBD properly transform the translational forces of the particles to translational movements of the mechanism.

4.3.1.4 Torsional Spring Damper System

The final test of the coupling of DEM and MBD method is performed by confirming the response of a torsional spring and damper system. Aim of this test is to verify that the MBD and DEM code are coupled correctly to the coupling server, both for translations and rotations. The motor of the pendulum example in Figure 4.15a has been replaced by a torsional spring. The system can then by described by

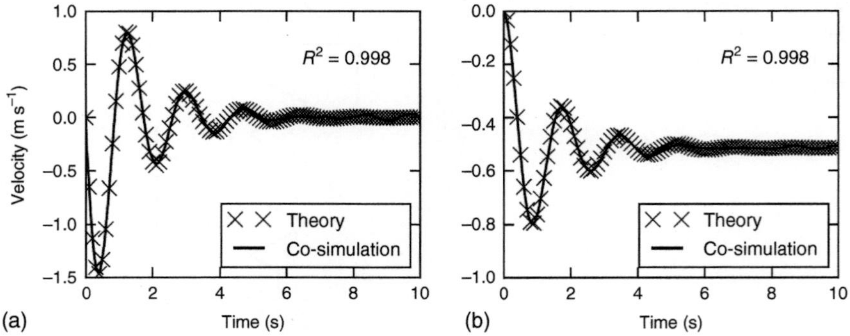

Figure 4.17 System response by the co-simulation compared to the analytical solution of Eq. (4.7) for a box with a mass of 100 kg, cumulative particle mass of 261.8 kg, spring constant k of 5000 N m^{-1}, preloaded at 100 N and the damper has a damping coefficient C_d of 500 Ns m^{-1}. Coefficient of restitution of the particles equals 1 (no damping). Source: Lommen et al. [44]/Emerald Publishing Limited/CC BY 4.0: (a) velocity of the box, and (b) position of the box.

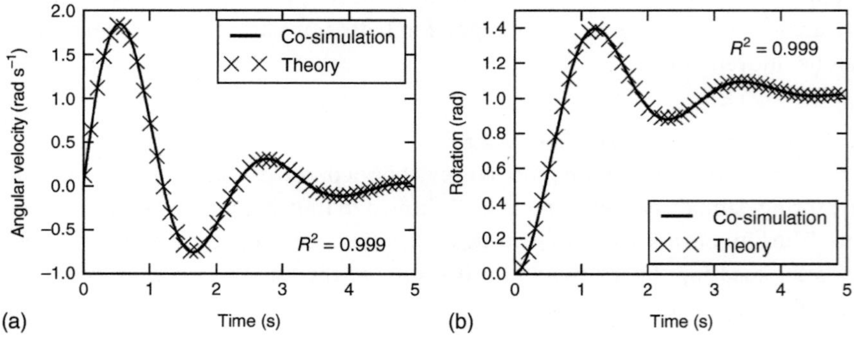

Figure 4.18 System response by the co-simulation compared to the analytical solution of Eq. (4.8) for a torsional spring stiffness k of 40 Nm deg^{-1} and a torsional damping coefficient C_d of 20 Nm deg^{-1}, mass of the box $m_c = 50$ kg, the total particle mass $m_p = 266.4$ kg. The initial position of the pendulum is horizontal. Source: Lommen et al. [44]/Emerald Publishing Limited/CC BY 4.0: (a) angular velocity and (b) rotation.

Eqs. (4.6) and (4.8):

$$\frac{d^2\theta}{dt^2}I + C_d\frac{d\theta}{dt} + k\theta = mgr\cos\theta \tag{4.8}$$

where I is the moment of inertia and the load on the system is determined by m, as the summation of the mass of the box and the total particle mass, and the angle of the pendulum θ.

In Figure 4.18, the angular velocity and rotation of the pendulum are shown. Due to the weight of the box with particles, the pendulum starts accelerating. The torsional spring prevents the pendulum from reaching the vertical position, whereas the torsional damper reduces the angular velocity. Comparing the simulation outcome to the equation of motion, it shows the coupling of DEM and MBD produces very accurate results with a determination coefficient of 0.999.

This final test concludes the verification process of the coupling method. Successful resemblance of the analytical solutions proves that the developed coupling of DEM and MBD works as expected and is capable of accurately simulating systems where particles and mechanisms interact.

4.3.2 Coupling Stability

Elliott [57] demonstrates several examples where the co-simulation results are not computed accurately, because of combination of a continuous MBD solver and an external discrete solver. The MBD solver selects its own timestep and requires only the error tolerance to be configured. However, each time a communication takes place, a timestep is forced in the MBD solver. The maximum MBD timestep Δt_{MBD} therefore depends on the DEM timestep Δt_{DEM} and the communication interval n (Eq. (4.9)):

$$\Delta t_{\mathrm{MBD}} \leq n \Delta t_{\mathrm{DEM}} \tag{4.9}$$

Besides an accurate co-simulation of DEM and MBD, a robust coupling is also desired. Both the solvers of the DEM and MBD have their own preferences for achieving a stable simulation with the lowest possible computational costs. However, when the two solvers are coupled in a co-simulation, these preferences sometimes conflict and need to be resolved.

For this reason, Lommen et al. [44] developed a guideline for achieving a stable and efficient co-simulation based on evaluating stability on a range of communication intervals n as defined in Eq. (4.9). They assessed the robustness of the coupling by evaluating stability at different communication intervals. A communication interval that is too large can be recognised by comparing the velocity profiles of the geometries to the derivative of the position data. An interval that is too small can be identified by examining the forces acting on the geometry. If these forces are extremely erratic, it is likely that the MBD will have difficulties in successfully computing the accompanying velocities and displacements.

4.4 How DEM–MBD Models Can be Used

Figure 4.19 shows a comprehensive approach to develop and use a DEM–MBD co-simulation for industrial applications. First, **DEM and MBD models** are developed separately in Steps A1, A2, and B1. Once adequate levels of **accuracy**, **stability** and **efficiency** in separate models are achieved, **co-simulation** (C1) can be set up using a **verified coupling framework**. In Step C2, the co-simulation is **validated** by using full-scale experiments in operational conditions. The validated co-simulation can be used by researchers and engineers to **optimise** existing solutions or to **innovate** new concepts. We describe essential examples and guidelines in this section on developing and using the DEM–MBD models.

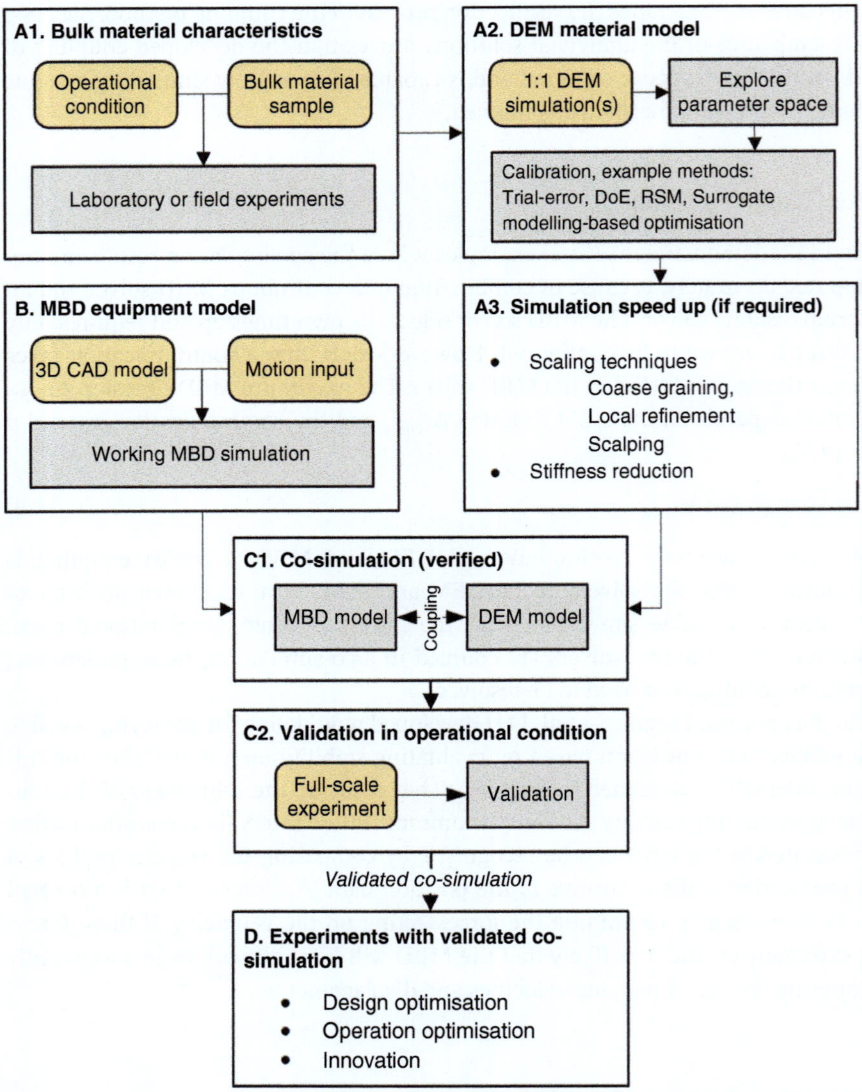

Figure 4.19 A comprehensive approach to develop and used DEM–MBD co-simulation for industrial applications.

4.4.1 DEM Calibration

According to [49], 'calibration is the estimation and adjustment of model parameters and constants to improve the agreement between model output and a data set'. Laboratory or field experiments are conducted to create a data set that includes bulk material characteristics. A comprehensive list of measurement methods for bulk characteristics can be found in [58]. In DEM calibration, bulk material characteristics are used to select appropriate simulation input parameters. In [36]

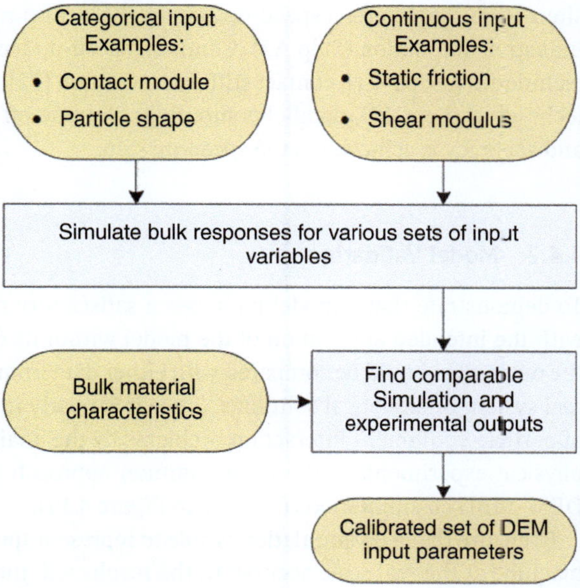

Figure 4.20 Main components of a generic DEM calibration procedure.

DEM simulation, input parameters are categorised in three different group of parameters:

(I) particle property,
(II) particle–geometry interaction, and
(III) particle–particle interaction.

Examples of particle properties are shape, size, and density. Coefficient of static friction, rolling friction and restitution are examples of particle–particle interaction parameters. To develop DEM–MBD co-simulations, it is essential to consider particle-geometry interaction parameters as well. Over the past decade, reliable DEM calibration procedures have been developed to model free-flowing bulk materials such as iron ore pellets [59], glass beads [60], sinter ore [61], sand [62, 63], and gravel [64, 65]. In addition, cohesive bulk materials such as moist iron ore fine [38, 66], coal [67], and powders [68] are modelled properly in DEM. A generic calibration approach is shown in Figure 4.20. Since various DEM simulation input parameters and bulk material characteristics play a role in the DEM calibration procedure, the trial-error approach is not recommended as it can be very time-consuming without a guarantee to reach a solution space. Systematic approaches such as Design of Experiments (DoE), Response Surface Method (RSM), and Surrogate modelling-based optimisation (SMO), and other proper optimisation techniques are recommended for DEM calibration.

In Step A1 in Figure 4.19, it is common to use laboratory-scale test methods to characterise bulk materials for DEM calibration. This is usually a time- and cost-effective approach compared to calibrating DEM input parameters using full-scale experiments. In general, a lower number of particles is required to be simulated in a laboratory-scale simulation as compared to a full-scale industrial simulation. Therefore, a speed-up step might be also required to set up a full-scale industrial simulation (Step A3). Common speed-up techniques are particle scaling techniques [59, 69–71], contact stiffness reduction [72], and efficient computational techniques (i.e. GPU, parallel computing). A trade-off between computation time and accuracy is achieved in the speed-up step.

4.4.2 Model Validation

To demonstrate that a model possesses a satisfactory range of accuracy consistent with the intended application of the model within its domain of applicability [49], the outcomes should be compared with either data from physical experiments of the real system or analytical solutions. Since it is nearly impossible to solve a majority of particle-equipment interaction problems by the analytical approach, conducting physical experiments is the most common approach to validate the results from DEM–MBD co-simulations (Step C2 in Figure 4.19).

To ensure that a co-simulation is able to represent the operating performance of a machine at the real scale accurately, the results will therefore have to be compared with full-scale *in situ* experiments. Conducting such experiments helps to capture the particle-equipment interaction in full-scale operational conditions. However, conducting a full-scale *in situ* experiment requires an operational piece of equipment, a considerable amount of bulk material, and the required measurement technologies to provide data for simulation validation. Reasons for not selecting full-scale validations include time and cost, lack of sensor and measurement technologies, and availability of equipment. As a result, the decision to conduct full-scale experiments depends on the research budget and goal.

An alternative approach is conducting laboratory-scale experiment, which represents operational conditions. Usually, scaling laws need to be employed to develop laboratory-scale experiments that may ignore some aspects of the process. There are a number of studies that have not validated the models in either full or laboratory scale and therefore only gain a global qualitative insight of machine performance.

In general, the full-scale validation requires experiments, simulation input data and statistical validation. Figure 4.21 demonstrates how the validation step is conducted for a grab co-simulation. In this example, experimental data includes crane operation and bulk material data. To set up the co-simulation, beside the experimental data, simulation input data such as 3D CAD model of equipment and bulk material model are required.

For a co-simulation of both particulate material and equipment, three software components are required. The first component is the DEM for simulation of particle behaviour and forces acting on the equipment. These loads are sent to the MBD component which consists of a mechanism of multiple bodies connected to each other through cables, joints, and contacts. The load on the mechanism affects the position of its bodies and these new positions are sent back to the DEM programme. The coupling consists of sending the position of the selected bodies from the dynamic model to the particle model and sending the load data of the selected

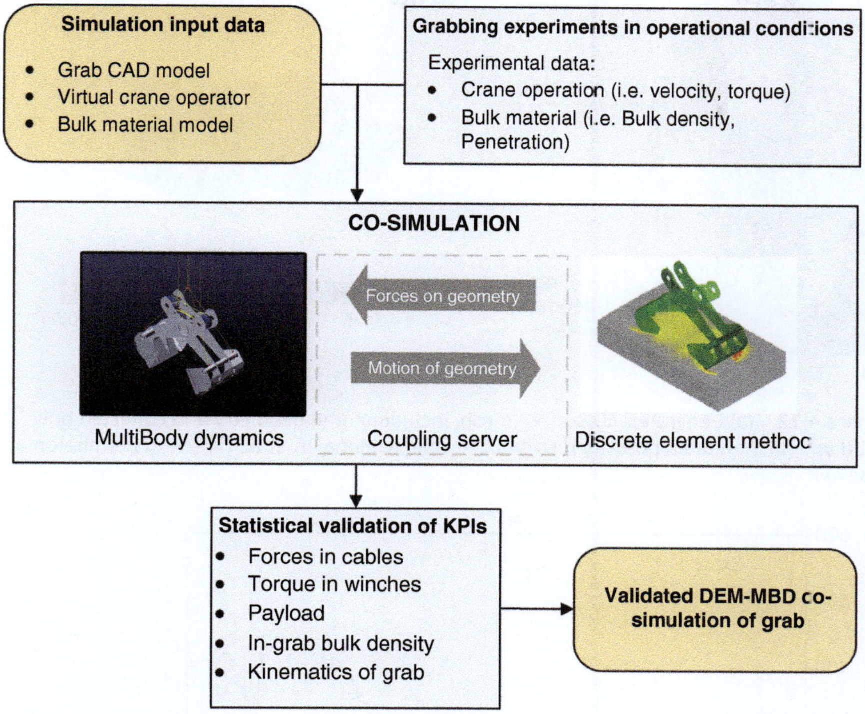

Figure 4.21 Overview of validating a DEM–MBD co-simulation for the grab application.

bodies from the particle model to the dynamic model. A coupling server, the third software component, is used to communicate between the two.

In the above example, various KPIs such as force, torque, and kinematics data are used to validate the co-simulation from multiple aspects. Therefore, proper measurement methods are needed to record the required data/signals during experiments. For example, a video-tracking method is used in [36] to determine grab kinematics during experiments. Also, the 3D-scanning technique is used in [37] to determine the volume and porosity of collected bulk material, as shown in Figure 4.22a,b respectively.

Statistical measures such as coefficient of determination (R^2) and Root mean square error (RMSE) are recommended to be used for confirming the accuracy of the signal-type data in the co-simulation. For example, Schott et al. [36] used R^2 to evaluate the accuracy of the total force in grab cables between the experiment and the co-simulation, as shown in Figure 4.23. The total force during the lowering stage (L) represents the empty weight of grab. Once the grab is resting on the bulk solid, cables become slack, and the total force drops. By tensing the closing cables during the next stage (C), the grab starts to collect the bulk solid. This continues by involving the hoisting cables and eventually suspending the grab out of the bulk material. The force over all stages is predicted accurately with an overall coefficient of determination of 0.958. By repeating the validation experiments, a high level of confidence in the validated co-simulation is gained.

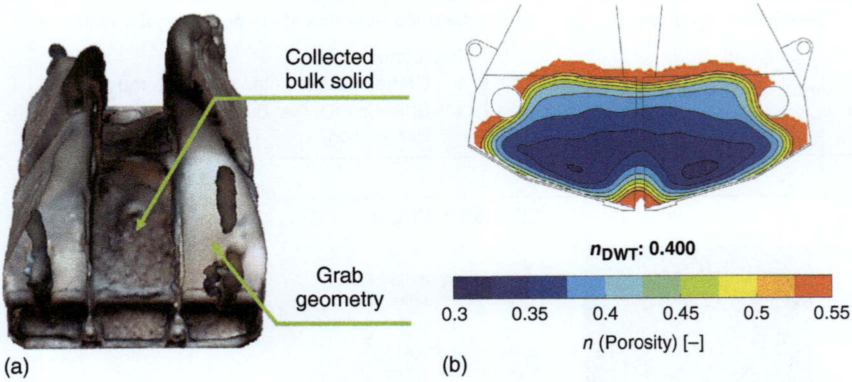

Figure 4.22 (a) Generated 3D surface mesh, including grab geometry and collected bulk solid and (b) Simulated porosity distribution. Source: Mohajeri et al. [37]/with permission of Elsevier.

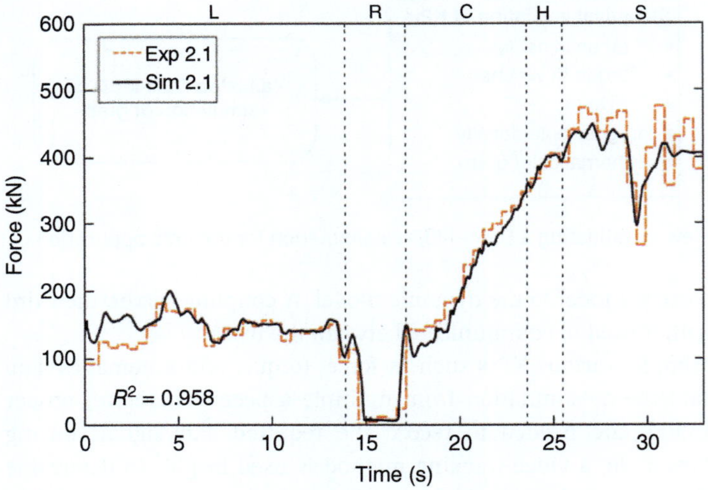

Figure 4.23 Load comparison between co-simulation and experiment for a grab. Grab operation consists of lowering of the grab (L), resting on the surface (R), closing (C), hoisting (H), and being suspended from the crane (S). Source: Schott et al. [36]/With permission of Elsevier.

4.4.3 Experimenting with a Validated Model

Once the co-simulation is validated, it can be used for experimenting methods to improve design and operation. The experiments with the validated model can be potentially beneficial to all parties concerned in the bulk handling process. The client's requirements (i.e. terminal or plant operator) can be considered in the design procedure to simulate a customised design/process. The designer can virtually prototype equipment in interaction with the bulk material within a reasonable time to include various bulk states, operational scenarios, and design configurations.

A major challenge in experimenting with validated model is to use resources (i.e. time, computation power) in an effective manner, so that a decisive conclusion can be derived. Therefore, a trial–error approach is not recommended as there is a risk that the high number of design and operational variables and their interactions could lead to inconclusive results. Systematic methods such as Design of Experiments and Surrogate modelling-based optimisation are proper tools for experimenting with a computationally expensive simulation [73]. First, the parameter space is explored using a Design of Experiment technique. Schott et al. [36] used Design of Experiment to explore various design configurations, which led to a breakthrough in ship unloading. Statistical tools such as main effect and interaction plots can be used to realise the influence of input variables on KPIs. Second, a response optimiser (e.g. meta-heuristic) in combination with a surrogate model can be used to optimise one or multiple KPIs jointly. A surrogate model is created by fitting a regression function to map the relationship between input variables (i.e. design and operational variables) and the selected response of the system (i.e. KPIs). Mohajeri et al. [74] used the surrogate modelling approach to enhance the grab design while considering multiple operational variables (e.g. bulk material states).

Also, experimenting with the validated co-simulation can be used to find bottlenecks in the operation and to explore ways for improvement. For example, de Vries [75] simulated a grab on various bulk surfaces to determine how the initial position of equipment influences the process KPIs. First, a grabbing process is simulated on a flat bulk surface (Figure 4.24a). The second cycle is simulated by moving the grab in a horizontal plane (Figure 4.24b). Therefore, the equipment performance in combination with an operator decision is evaluated when the bulk surface is irregular.

Furthermore, simulation-supported innovation is a promising approach to design novel bulk handling mechanisms and systems. The primary difference between optimisation and innovation is that in optimisation, available resources (e.g. existing equipment) is used in an effective manner to maximise KPIs, while in innovation, a new concept/solution is developed that did not exist before. For example, one can innovate a new grab concept for unloading bulk materials from the cargo hold.

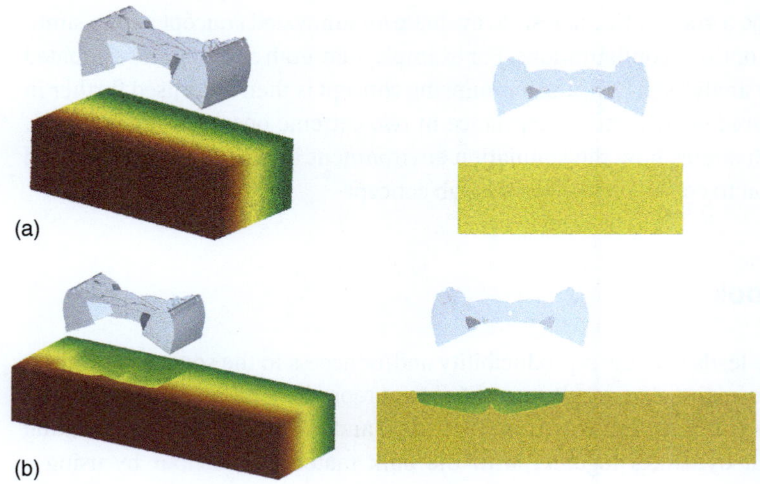

Figure 4.24 Simulating the grabbing process on flat (a) and irregular bulk surfaces (b).

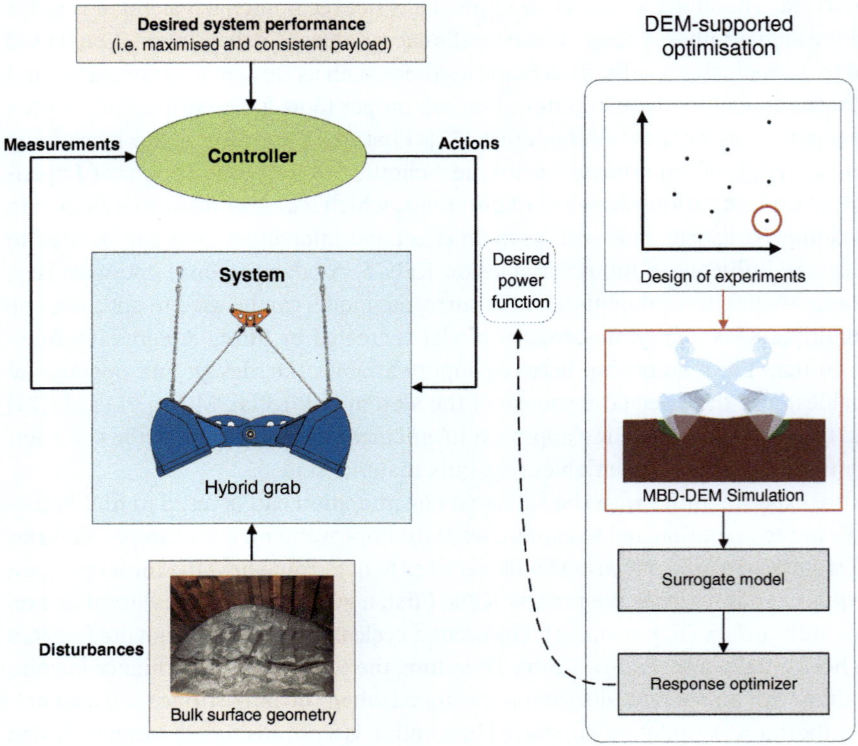

Figure 4.25 Hybrid grab system: creating a variable and controllable power function for desired grab performance under varying operational conditions.

The performance of innovated concepts can be evaluated in a DEM–MBD simulation environment. Although innovation can be quick, powerful, and effective, there is a risk regarding unpredictability and uncertainty of new concepts. In other words, there should be a systematic analysis to evaluate an innovated concept and to simulate it with its optimal configurations. For example, new grab concepts are evaluated in a feasibility analysis in [76], and a promising concept is then optimised further in terms of required energy and performance in two extreme operational conditions. Figure 4.25 illustrates how the simulation environment is used to find the desired power function to control a new hybrid grab concept.

4.5 Outlook

This therefore leads to a low reproducibility and/or access to the source codes from a scientific viewpoint. The results are not always reproducible. Scientific literature has shown the value of DEM in equipment design and specifically for incorporating the equipment dynamics together with the bulk material behaviour by using a coupled DEM–MBD approach. Insights in performance are gained and used in the design process. The reported results include increase in payloads, decrease in masses and number of parts altogether reducing manufacturing costs tremendously [36]. Although a few studies have been shown to use the validated models to explore design space and operational performance there is still a huge potential for industry. This can only be fully exploited when models are systematically calibrated and validated.

Especially with the development in the increase in computational power, more realistic particle shapes can be seriously considered, or we could move towards system level simulations such as a crane unloading an entire cargo hold, or a scraper reclaiming material in an enclosed stockpile. However, an increase in computational capabilities also means that more complex simulations and interactions will be modelled. This will eventually result in still quite long simulation times, likely similar to what we find acceptable today.

Since important environmental parameters such as the gravitational constant can be changed, DEM–MBD offers the possibility to conduct simulations that take the environment on other planets into account. It is therefore expected that DEM–MBD will increasingly play a role in space expedition simulation. However, the importance of material calibration remains equally important to be able to develop models that accurately resemble soils, minerals, or granular materials and the interaction with the equipment.

Other potential is the use of exploring advanced control strategies using the coupling towards the development from digital shadows towards digital twins. The digital twin can interact with real-world physical and allow for two-way data transmission. This also requires the development of advanced sensors capable of measuring *in situ* bulk behaviour, equipment and process performance, interaction between material and equipment interfaces in the right time horizons. For operational control, these time horizons are real-time or near real-time, while for design optimisation, much longer time horizons are applicable, typically hours or even days.

References

1 Lommen, S. W. (2016). Virtual prototyping of grabs: co-simulations of discrete element and rigid body models. Delft University of Technology, Delft.

2 Mohajeri, M., (2021). Grabs and cohesive bulk solids: virtual prototyping using a validated co-simulation. PhD thesis, TU Delft. https://doi.org/10.4233/uuid:b232e542-4881-4b02-8677-a7b1dd37b6b0

3 Cundall, P.A. and Strack, O.D.L. (1979). A discrete numerical model for granular assemblies. *Geotechnique* 29 (1): 47–65.

4 Langerholc, M., Česnik, M., Slavič, J., and Boltežar, M. (2012). Experimental validation of a complex, large-scale, rigid-body mechanism. *Engineering Structures* 36: 220–227.

5 Whittaker, E. T. (1937). A treatise on the analytical dynamics of particles and rigid bodies. CUP Archive.

6 Wittenburg, J. (2013). *Dynamics of Systems of Rigid Bodies*, vol. 33. Springer-Verlag.

7 Coetzee, C.J., Els, D.N.J., and Dymond, G.F. (2010). Discrete element parameter calibration and the modelling of dragline bucket filling. *Journal of Terramechanics* 47 (1): 33–44. https://doi.org/10.1016/j.jterra.2009.03.003.

8 Li, X., Duan, C., Bai, K., and Yao, Z. (2021). Operating performance of pure electric loaders with different types of motors based on simulation analysis. *Energies* 14 (3): 1–19. https://doi.org/10.3390/en14030617.

9 Edwards, B. (2018). Co-simulation of mbd models with dem code to predict mobility on soft soil. In: *Proceedings of the Ground Vehicle Systems Engineering and Technology Symposium (GVSETS)*, 1–14. Novi, MI: NDIA.

10 E. Pesheck, T. Palmer, T. Bromwell, and V. Jeganathan, "Proceedings of the Ground Vehicle Systems Engineering and Technology Symposium (GVSETS),"NDIA, Novi, MI, 2019.

11 Shin, Y.J., Jeong, J.S., Jun, C.W., and Sohn, J.H. (2020). Interacting analysis between wheel and sand particles based on DEM and its validation with experiments. *Journal of Mechanical Science and Technology* 34 (11): 4537–4544. https://doi.org/10.1007/s12206-020-1013-7.

12 Teslenko, D.S. and Belyakov, V.V. (2021). Simulation of the process of interaction between the vehicle and snow in the environment of DEM and MBD-systems. *IOP Conference Series: Materials Science and Engineering* 1086 (1): 012036. https://doi.org/10.1088/1757-899x/1086/1/012036.

13 Hongyue, C., Shihao, Y., Huming, N. et al. (2021). Analysis of the driving and passing performance of bolter miner under different geological conditions. *Journal of China Coal Society* 46 (7): 2112–2122.

14 Nakashima, H. et al. (2010). Discrete element method analysis of single wheel performance for a small lunar rover on sloped terrain. *Journal of Terramechanics* 47 (5): 307–321. https://doi.org/10.1016/j.jterra.2010.04.001.

15 Thoesen, A., McBryan, T., and Marvi, H. (2019). Helically-driven granular mobility and gravity-variant scaling relations. *RSC Advances* 9 (22): 12572–12579. https://doi.org/10.1039/c9ra00399a.

16 Thoesen, A., McBryan, T., Mick, D. et al. (2020). Comparative performance of granular scaling laws for lightweight grouser wheels in sand and lunar simulant. *Powder Technology* 373: 336–346. https://doi.org/10.1016/j.powtec.2020.05.114.

17 Ji, S. and Liang, S. (2021). DEM-FEM-MBD coupling analysis of landing process of lunar lander considering landing mode and buffering mechanism. *Advances in Space Research.* 68 (3): 1627–1643. https://doi.org/10.1016/j.asr.2021.03.034.

18 Liu, G. et al. (2021). Study on structural health monitoring of vertical vibration of ballasted track in high-speed railway. *Journal of Civil Structural Health Monitoring* 11 (2): 451–463. https://doi.org/10.1007/s13349-020-00460-x.

19 Dai, Y., Yin, W., and Ma, F. (2019). Nonlinear multi-body dynamic modeling and coordinated motion control simulation of deep-sea mining system. *IEEE Access* 7: 86242–86251. https://doi.org/10.1109/ACCESS.2019.2925714.

20 Dai, Y., Li, X., Yin, W. et al. (2021). Dynamic modelling and motion control research on deep seabed miner. *Marine Georesources & Geotechnology* 39 (4): 389–397. https://doi.org/10.1080/1064119X.2019.1706675.

21 Thoeni, K., Servin, M., Sloan, S.W., and Giacomini, A. (2019). Designing waste rock barriers by advanced numerical modelling. *Journal of Rock Mechanics and Geotechnical Engineering* 11 (3): 659–675. https://doi.org/10.1016/j.jrmge .2018.11.005.

22 Ghoohaki, S. A. (2013). High-fidelity modeling of a backhoe digging operation using an explicit multibody dynamics finite element code with integrated discrete element method. Purdue University.

23 Sane, A. G. (2015). High-fidelity modelling of a bulldozer using an explicit multibody dynamics finite element code with integrated discrete element method. Purdue University. https://doi.org/10.7912/C2/2606

24 Axås, J. G. (2020). A coupled multibody and discrete element approach for roller compaction dynamics. Chalmers University of Technology. https://hdl.handle .net/20.500.12380/301057.

25 Shi, S., Gao, L., Cai, X. et al. (2020). Effect of tamping operation on mechanical qualities of ballast bed based on DEM-MBD coupling method. *Computers and Geotechnics* 124 (March): 103574. https://doi.org/10.1016/j.compgeo.2020.103574.

26 Shi, S., Gao, L., Xiao, H. et al. (2021). Research on ballast breakage under tamping operation based on DEM–MBD coupling approach. *Construction and Building Materials* 272: 121810. https://doi.org/10.1016/j.conbuildmat.2020.121810.

27 Lichtenheldt, R., Schäfer, B., and Krömer, O. (2014). Hammering beneath the surface of Mars-Modeling and simulation of the impact-driven locomotion of the HP3-Mole by coupling enhanced multi-body dynamics. In: *58th Ilmenau Scientific Colloquium*, no. July, 1–20. https://d-nb.info/1069679488/34.

28 Alkalla, M., Pang, X., Pitcher, C., and Gao, Y. (2021). DROD: a hybrid biomimetic undulatory and reciprocatory drill: quantitative analysis and numerical study. *Acta Astronautica* 182 (February): 131–143. https://doi.org/10.1016/j .actaastro.2021.02.007.

29 Cheng, J., Ren, T., Zhang, Z. et al. (2020). A dynamic model of inertia cone crusher using the discrete element method and multi-body dynamics coupling. *Minerals* 10 (10): 1–23. https://doi.org/10.3390/min10100862.

30 Cheng, J., Ren, T., Zhang, Z. et al. (2021). Influence of two mass variables on inertia cone crusher performance and optimization of dynamic balance. *Minerals* 11 (2): 1–18. https://doi.org/10.3390/min11020163.

31 Barrios, G.K.P. and Tavares, L.M. (2016). A preliminary model of high pressure roll grinding using the discrete element method and multi-body dynamics coupling. *International Journal of Mineral Processing* 156: 32–42. https://doi.org/10 .1016/j.minpro.2016.06.009.

32 Rodriguez, V.A., Barrios, G.K.P., Bueno, G., and Tavares, L.M. (2021). Investigation of lateral confinement, roller aspect ratio and wear condition on HPGR performance using DEM-MBD-PRM simulations. *Minerals* 11 (8): https://doi.org/ 10.3390/min11080801.

33 Wang, Y., Yu, J., Yu, Y., and Zhang, Q. (2019). Validation of a coupled model of discrete element method with multibody kinematics to simulate the screening process of a swing-bar sieve. *Powder Technology* 346: 193–202. https://doi org/10 .1016/j.powtec.2019.01.076.

34 Wu, B., Zhang, X., Niu, L. et al. (2019). Research on sieving performance of flip-flow screen using two-way particles-screen panels coupling strategy. *IEEE Access* 7: 124461–124473. https://doi.org/10.1109/ACCESS.2019.2938847.

35 Hess, G., Richter, C., and Katterfeld, A. (2016). Simulation of the dynamic interaction between bulk material and heavy equipment: calibration and validation. In: *The 12th International Conference on Bulk Materials Storage, Handling and Transportation (ICBMH)*, 427–436. Engineers Australia, Barton: ACT.

36 Schott, D.L., Mohajeri, J., Jovanova, J. et al. (2021). Design framework for DEM-supported prototyping of grabs including full-scale validation. *Journal of Terramechanics* 96: 29–43. https://doi.org/10.1016/j.jterra.2021.04.003.

37 Mohajeri, M.J., de Kluijver, W., Helmons, R.L.J., van Rhee, C., Schott, D.L. (2021). A validated co-simulation of grab and moist iron ore cargo: replicating the cohesive and stress-history dependent behaviour of bulk solids. *Advanced Powder Technology*. https://doi.org/10.1016/j.apt.2021.02.017.

38 Mohajeri, M.J., van Rhee, C., and Schott, D.L. (2021). Replicating stress-history dependent behaviour of cohesive solid materials: feasibility and definiteness in DEM calibration procedure. *Advanced Powder Technology*. https://doi.org/10.1016/j.apt.2021.02.044.

39 Fimbinger, E. and Becker, A. (2021). Multi-Physics DEM Simulation: Capabilities & features of the simulation software ThreeParticle/CAE as a powerful platform for virtual prototyping with complex bulk materials. In: *Projektbroschüre der Bergbaukunde 2020*, 32. University of Leoben.

40 Dan, D., Wang, Q., and Gong, J. (2021). Application of coupled multi-body dynamics—discrete element method for optimization of particle damper for cable vibration attenuation. *Frontiers of Structural and Civil Engineering* 15 (1): 244–252. https://doi.org/10.1007/s11709-021-0696-x.

41 Kobayashi, T., Ishigami, G., Ozaki, S., et al. Current status and prospects of terramechanics-based simulation techniques for planetary rover locomotion. The International Symposium on Artificial Intelligence, Robotics and Automation in Space (i-SAIRAS 2010) August 29 - September 1, 2010 Sapporo Convention Center, Sapporo, Japan.

42 Smith, W., Melanz, D., Senatore, C. et al. (2013). Comparison of DEM and traditional modeling methods for simulating steady-state wheel-terrain interaction for small vehicles. In: *Proceedings of the 7th Americas Regional Conference of the ISTVS*, 4–7. Tampa: FL, USA.

43 Chen, Z., Xue, D., Wang, G. et al. (2021). Simulation and optimization of the tracked chassis performance of electric shovel based on DEM-MBD. *Powder Technology* 390: 428–441. https://doi.org/10.1016/j.powtec.2021.05.085.

44 Lommen, S., Lodewijks, G., and Schott, D.L. (2018). Co-simulation framework of discrete element method and multibody dynamics models. *Engineering Computations* 34: 1481–1499.

45 Shmulevich, I. (2010). State of the art modeling of soil-tillage interaction using discrete element method. *Soil and Tillage Research* 111 (1): 41–53. https://doi.org/10.1016/j.still.2010.08.003.

46 Lommen, S.W., Schott, D.L., and Lodewijks, G. (2012). Multibody dynamics model of a scissors grab for co-simulation with discrete element method. *FME Transactions* 40 (4): 177–180.

47 Chung, Y.C. and Wu, Y.R. (2019). Dynamic modeling of a gear transmission system containing damping particles using coupled multi-body dynamics and discrete element method. *Nonlinear Dynamics* 98 (1): 129–149. https://doi.org/10.1007/s11071-019-05177-1.

48 Wu, Y.R., Chung, Y.C., and Wang, I.C. (2021). Two-way coupled MBD–DEM modeling and experimental validation for the dynamic response of mechanisms containing damping particles. *Mechanism and Machine Theory* 159: 104257. https://doi.org/10.1016/j.mechmachtheory.2021.104257.

49 Rykiel, E.J. Jr. (1996). Testing ecological models: the meaning of validation. *Ecological Modelling* 90 (3): 229–244.

50 Elmqvist, H., Goteman, A., Roxling, V., and Ghandriz, T. (2015, https://doi.org/10.3384/ecp15118427). Generic modelica framework for multibody contacts and discrete element method. In: *Proceedings of the 11th International Modelica Conference, Versailles, France, September 21-23, 2015*, vol. 118, no. September 2015, 427–440. Modelica Association and Linköping University Electronic Press.

51 Chen, W., van Etten, M., Donohue, T., and Williams, K. (2017). Application of the coupled discrete element modelling and modelica based multi-body dynamics in system-level modelling. In: *Proceedings of the 7th International Conference on Discrete Element Methods. DEM 2016. Springer Proceedings in Physics*, vol. 188 (ed. X. Li, Y. Feng and G. Mustoe), 571–578. Singapore: Springer . https://doi.org/10.1007/978-981-10-1926-5_59.

52 Richter, C., Roessler, T., Otto, H., and Katterfeld, A. (2021). Coupled discrete element and multibody simulation, part I: implementation, verification and validation. *Powder Technology* 379: 494–504. https://doi.org/10.1016/j.powtec.2020.10.074.

53 Yoon, J.S., Park, J.S., Ahn, C.O., and Choi, J.H. (2011). Cosimulation of MBD (multi body dynamics) and dem of many spheres using GPU technology. *Part. Methods II – Fundamentals and Applications* 778–785.

54 Ortiz, J.L. and MacDonald, N. (2016). Co-simulation of large mbd, FEA and DEM system. In: *ECCOMAS Congress 2016 - Proceedings of the 7th European Congress on Computational Methods in Applied Sciences and Engineering*, vol. 1, no. May, 1832–1837. National Technical University of Athens (NTUA). doi: 10.7712/100016.1921.11733.

55 Sanborn, G., Choi, J., and Choi, J.H. (2021). Strategy for co-simulation of multi-flexible-body dynamics and the discrete element method. *Journal of Mechanical Science and Technology* 35 (10): 4363–4380. https://doi.org/10.1007/s12206-021-0908-2.

56 Weisberg, S. (2005). *Applied Linear Regression*, vol. 528. Wiley.

57 Elliott, A. (2000). A Highly Efficient, General-Purpose Approach for Co-Simulation with ADAMS®.

58 Schulze, D. (2008). *Powders and Bulk Solids – Behaviour, Characterization, Storage and Flow*. Springer.

59 Lommen, S., Mohajeri, M., Lodewijks, G., and Schott, D.L. (2019). DEM particle upscaling for large-scale bulk handling equipment and material interaction. *Powder Technology* 352: 273–282. https://doi.org/10.1016/j.powtec.2019.04.034.

60 Rackl, M. and Hanley, K.J. (2017). A methodical calibration procedure for discrete element models. *Powder Technology* 307: 73–83. https://doi.org/10.1016/j.powtec.2016.11.048.

61 Benvenuti, L., Kloss, C., and Pirker, S. (2016). Identification of DEM simulation parameters by Artificial Neural Networks and bulk experiments. *Powder Technology* 291: 456–465. https://doi.org/10.1016/j.powtec.2016.01.003.

62 Derakhshani, S.M., Schott, D.L., and Lodewijks, G. (2015). Micro–macro properties of quartz sand: experimental investigation and DEM simulation. *Powder Technology* 269: 127–138. https://doi.org/10.1016/j.powtec.2014.08.072.

63 Do, H.Q., Aragón, A.M., and Schott, D.L. (2018). A calibration framework for discrete element model parameters using genetic algorithms. *Advanced Powder Technology* 29 (6): 1393–1403. https://doi.org/10.1016/j.apt.2018.03.001.

64 Richter, C., Rößler, T., Kunze, G. et al. (2019). Development of a standard calibration procedure for the DEM parameters of cohesionless bulk materials – Part II: Efficient optimization-based calibration. *Powder Technology* 360: 967–976. https://doi.org/10.1016/j.powtec.2019.10.052.

65 Roessler, T., Richter, C., Katterfeld, A., and Will, F. (2019). Development of a standard calibration procedure for the DEM parameters of cohesionless bulk materials–part I: solving the problem of ambiguous parameter combinations. *Powder Technology* 343: 803–812.

66 Morrissey, J. P. (2013). Discrete element modelling of iron ore pellets to include the effects of moisture and fines. PhD Thesis.

67 Mohajeri, M.J., Do, H.Q., and Schott, D.L. (May 2020). DEM calibration of cohesive material in the ring shear test by applying a genetic algorithm framework. *Advanced Powder Technology* 31: 112–125. doi: 10.1016/j.apt.2020.02.019.

68 Orefice, L. and Khinast, J.G. (2019). A novel framework for a rational, fully-automatised calibration routine for DEM models of cohesive powders. *Powder Technology* 687–703.

69 Mohajeri, M.J., Helmons, R.L.J., van Rhee, C., and Schott, D.L. (2020). A hybrid particle-geometric scaling approach for elasto-plastic adhesive DEM contact models. *Powder Technology* 369: 72–87. https://doi.org/10.1016/j.powtec.2020.05.012.

70 Thakur, S.C., Ooi, J.Y., and Ahmadian, H. (2016). Scaling of discrete element model parameters for cohesionless and cohesive solid. *Powder Technology* 293: 130–137.

71 Roessler, T. and Katterfeld, A. (2018). Scaling of the angle of repose test and its influence on the calibration of DEM parameters using upscaled particles. *Powder Technology* 330: 58–66. https://doi.org/10.1016/j.powtec.2018.01.044.

72 Lommen, S., Schott, D., and Lodewijks, G. (2014). Particuology DEM speedup: stiffness effects on behavior of bulk material. *Particuology* 12: 107–112. https://doi.org/10.1016/j.partic.2013.03.006.

73 Bliek, L. (2022). A survey on sustainable surrogate-based optimisation. *Sustainability* 14 (7): 3867.

74 Mohajeri, M.J., van den Bergh, A.J., Jovanova, J., and Schott, D.L. (2021). Systematic design optimization of grabs considering bulk cargo variability. *Advanced Powder Technology* 32 (5): 1723–1734. https://doi.org/10.1016/j.apt.2021.03.027.

75 de Vries, N. (2020). Design of grabs under operational conditions.

76 van Kol, T. (2021). Hybrid grab concepts for varying operational conditions.

5

Simulation of Pneumatic Conveying

Don McGlinchey

Glasgow Caledonian University, Glasgow G4 0BA, Scotland, UK

5.1 Introduction

Pneumatic transport or pneumatic conveying, although a well-established industrial bulk solid handling operation with a long history, can still present many challenges both in the design of new conveying systems and troubleshooting existing systems. In the design context usually the (average) flow rate of solids and the location of the feed point and end point are known, and the pipeline route, system components and type, air (gas) requirements (flow rate and available pressure head), etc. need to be established. In troubleshooting, common problems are failure to meet the required solids throughput, pipeline blockages, excessive pipeline wear, and product degradation. Simulation at various scales can help to find and optimize solutions to these challenges. A pneumatic conveying system can be considered as a set of connected and interdependent zones; see Figure 5.1.

A pneumatic conveying system can be conceptualized as comprising of three zones in which both gas (usually air) and particulate solids are present and interacting.

Zone 1: Mixing of solids and gas; rotary valve plus dropout box, blow tank, screw feeder plus dropout box (vacuum system).
Zone 2: Pneumatic conveying pipeline; horizontal, vertical inclined, bends.
Zone 3: Gas–solids separation; cyclone separator

5.1.1 Pneumatic Conveying – Modes of Flow

The conveying of particulate solids in Zone 2 and the interaction and behaviour of the gas–solids flow can be described by 'modes' which are dependent on gas velocity, particle size, shape and density, solids volume fraction, etc. One of the most common classifications is based on the average particle concentration in the pipeline. Traditionally, researchers have categorised these modes of conveying

Simulations in Bulk Solids Handling: Applications of DEM and other Methods, First Edition.
Edited by Don McGlinchey.

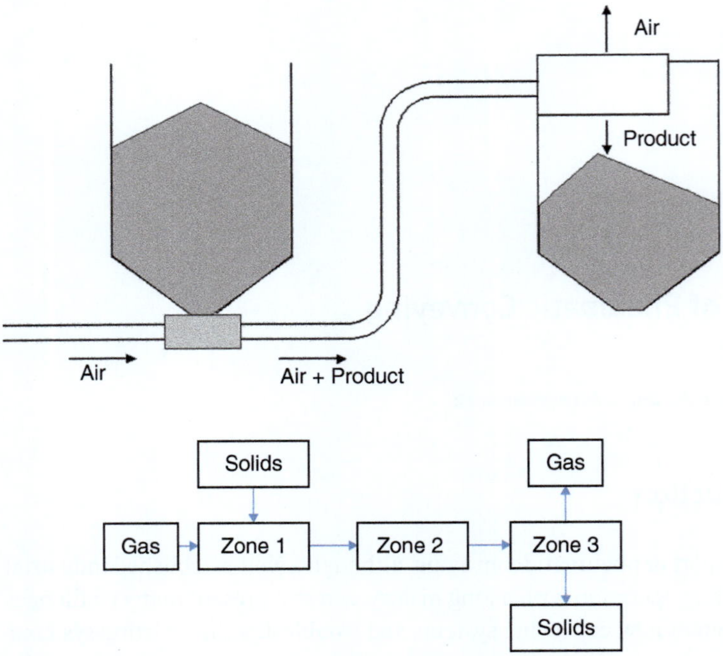

Figure 5.1 Outline of a pneumatic conveying system.

based on average particle concentration in the pipeline or by the ratio of product mass conveyed ($kg\,s^{-1}$) to the mass flow rate of air used for the conveying ($kg\,s^{-1}$). This is referred to as the solids loading ratio (–) [1–4].

Many of the modes are transient and change within the pipeline over time and location even if the system is operating at a nominal steady state. Three of the principle modes, namely dilute or suspension flow, plug flow, and moving bed flow, are shown in a simplified form in Figure 5.2.

5.1.2 Definition of Pneumatic Conveying Terms

5.1.2.1 Solids Loading Ratio

A commonly used parameter in pneumatic conveying design is the solids loading ratio (SLR) and is simply the ratio of the mass flow rate of solids to the mass flow rate of gas.

$$\mathrm{SLR} = \phi = \frac{\dot{m}_s}{\dot{m}_g} \tag{5.1}$$

5.1.2.2 Density Ratio

The density ratio (DR) is the ratio of solids density to gas density and in common pneumatic conveying situations will be of the order of 500–1000.

$$\mathrm{DR} = \frac{\rho_s}{\rho_g} \tag{5.2}$$

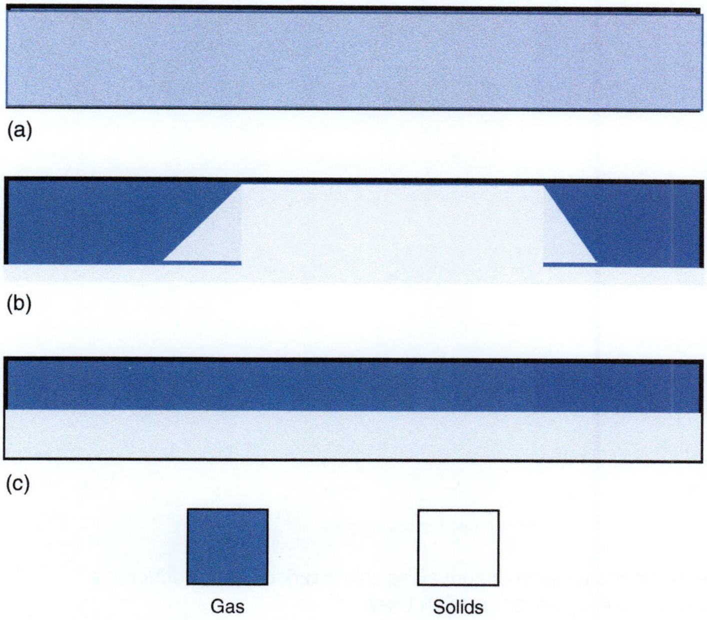

Figure 5.2 Simplified illustration of pneumatic conveying modes. (a) Dilute phase; fine particles suspended in gas flow. (b) Dense phase 1 Plug flow; large particles move as full bore slugs. (c) Dense phase 2 Moving bed; fine particles in a fluidized state.

5.1.2.3 Volume Fraction

The volume fraction (VF) is related to both the SLR and the DR and is the ratio of the volume occupied by solids to the volume occupied by the gas.

$$\mathrm{VF} = \alpha = \frac{V_\mathrm{s}}{V_\mathrm{g}} \tag{5.3}$$

5.1.2.4 Conveying Characteristics

The relationship between, gas mass flow rate, solids mass flow rate, and the line pressure drop can be summarised by a Performance Map or a Conveying Characteristic, for example Figure 5.3. The mass flow rate of gas is often chosen rather than a volumetric flow rate as the mass flow rate is a constant along the length of the conveying line, whereas the volumetric flow rate increases along the length of the line towards the exit. The system operating point on such a diagram can be used as a guide to the most appropriate simulation method. If the system operating point is in an area of the conveying characteristic with a high gas mass flow rate and low solids flow rate i.e. with a low SLR and can be characterized as a dilute phase, then several simulation methods can be utilized, for example, DPM, TF, and computational fluid dynamics–discrete element method (CFD–DEM), as appropriate for the study. Conversely, if the operating point is in a dense phase regime, then simulation choices are restricted, e.g. CFD–DEM or granular kinetic theory (GKT), depending on the nature of the material and the requirements of the study.

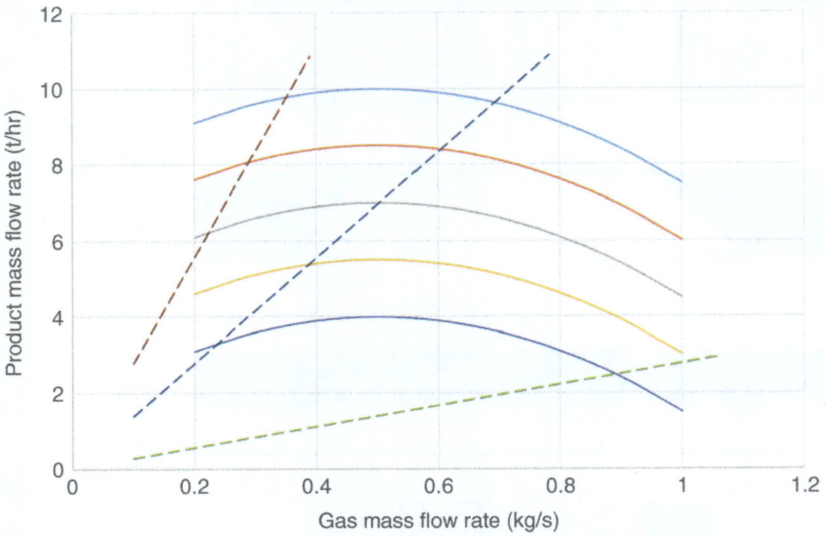

Figure 5.3 Example performance map or conveying characteristic of a product in a particular conveying line showing isobar and SLR lines.

Model Choices Can be Based on Gas–Solids–System Interactions

1. Influence of gas flow field on particle motion
2. Influence of the presence of particles on the gas flow field.
3. Particle–system solid boundary interaction.
4. Particle–particle collisions

5.1.3 Pressure Drop Prediction Versus Material Behaviour

There are typically two reasons to make a simulation of a pneumatic conveying system or a part of the system.

1. Pressure drop prediction along the length of the conveying line or across a pipeline component
2. Predict material flow behaviour, e.g. at a pipeline bend.

These two desired outcomes can occasionally be met by the same simulation, but this is often not the most efficient or practical approach.

1. A first approximation of the pressure drop along the length of a conveying line can be made with a relatively simple one-dimensional CFD model with additional terms to account for the presence of the particulate solids.
2. A study of the flow behaviour of a slug or moving bed of solids at a bend would require a three-dimensional model which could provide a realistic representation of both the fluid and the solids and the interaction between them (Figures 5.4–5.6).

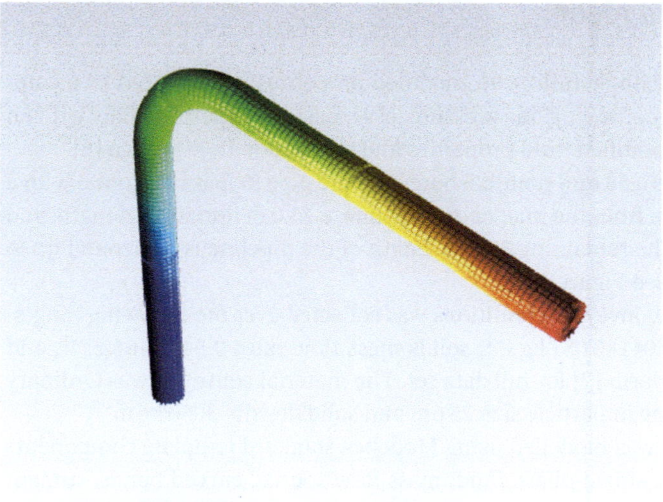

Figure 5.4 Example of pressure contour plot.

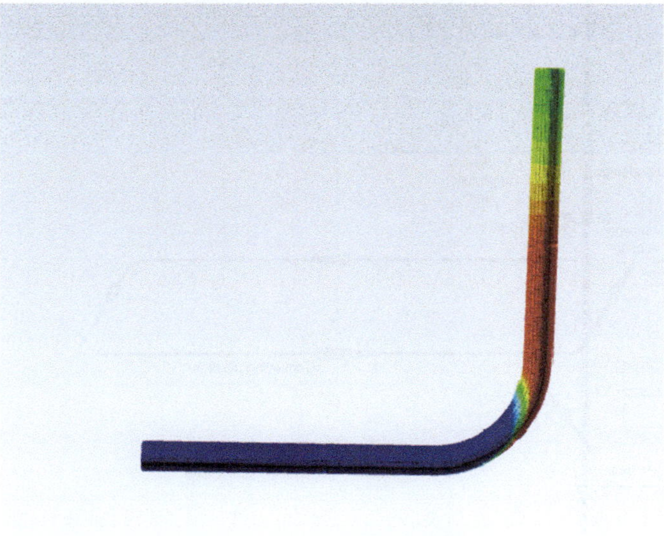

Figure 5.5 Example of Eulerian volume fraction plot of a plug entering a bend.

Figure 5.6 CFD–DEM simulation of one plug in the pipeline with particles of different particle density [5].

5.2 Single Fluid Model

An approach to use a single fluid with modified properties is illustrated by a comparison of pneumatic conveying line pressure loss measured experimentally with an estimation by using modified fluid properties and simulation in Modelica [6].

The pipeline used is a 53 mm nominal bore, and the pipe loop is horizontal with a 3.0 m vertical lift 5.0 m from the inlet of the pipeline, a 25.0 m horizontal length, and a vertical lift 0.5 m. The remaining 30.0 m length of the pipeline is horizontal up to the receiving vessel (see Figure 5.7).

A set of data for 40 conveying conditions was collected over the following ranges: air mass flow rates 0.041–0.081 kg s^{-1}; solids mass flow rates 0.54–2.46 kg s^{-1}; and SLRs 6.7–58.1. See Folarin [7] for full data set. The material conveyed was Ordinary Portland Cement of mean particle size 25 μm and solid density 3060 kg m^{-3}.

The conveying line was modelled using Modelica standard template components for an incompressible single-phase fluid; mass flow source, curved bends, straight pipe sections, and a pressure sink at the exit. The configuration developed in OpenModelica, as shown in Figure 5.8.

The pressure loss in pipeline sections is calculated using a D'Arcy type approach.

$$dp = \lambda(Re, \Delta) * \left(\frac{L}{D}\right) * \rho * v * |v| /2 \tag{5.4}$$

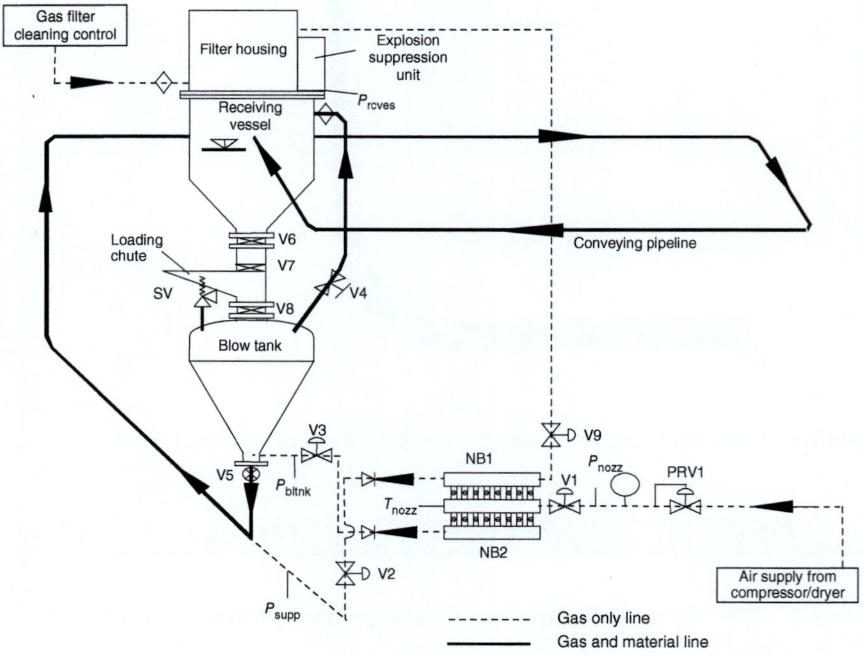

Figure 5.7 Schematic diagram of the pneumatic conveying test Rig at Glasgow Caledonian University.

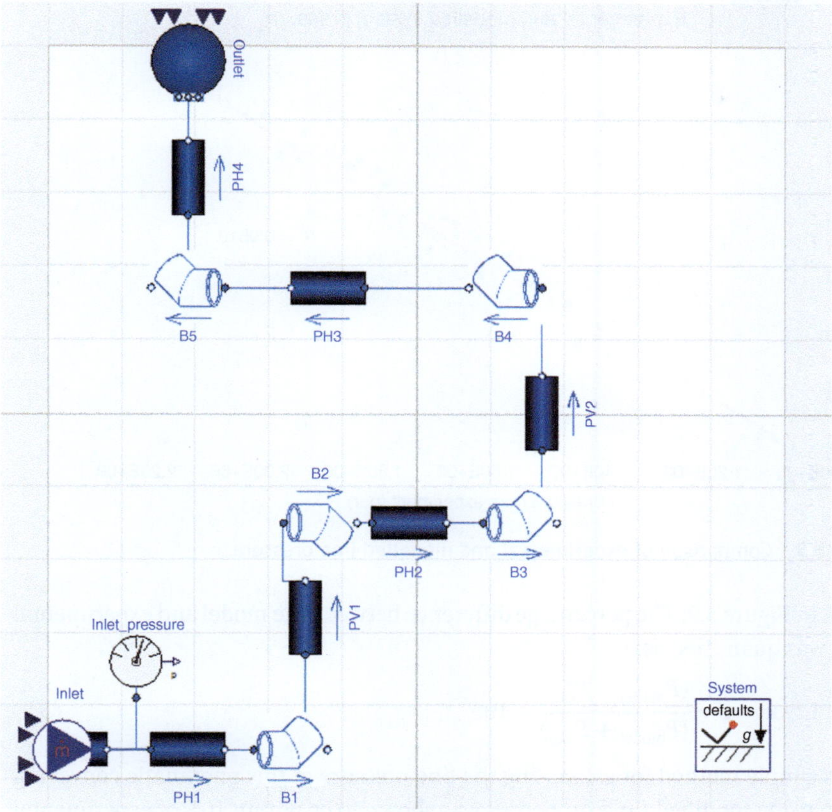

Figure 5.8 OpenModelica schematic of pipeline.

where the symbols have their usual meaning, and the friction factor lambda is calculated using the Colebrook-White equation approximation,

$$\frac{1}{\sqrt{\lambda}} = -2 * \log\left(\frac{2.51}{Re * \sqrt{\lambda}} + 0.27 * \Delta\right) \tag{5.5}$$

The pressure loss due to curved bends is similar to the calculation for straight pipes, as detailed in the OpenModelica documentation (http://openModelica.org).

The fluid properties from the gas–solids flow with a volume ratio VR were modified to a single phase using the following equation for a mean density ρ_{m}

$$\rho_{\mathrm{m}} = \rho_{\mathrm{a}} * (1 - \mathrm{VR}) + \rho_{\mathrm{s}} * \mathrm{VR} \tag{5.6}$$

And a modified viscosity

$$\mu_{\mathrm{m}} = \mu_{\mathrm{a}} * (1 + 2.5 * \mathrm{VR} + 10.05 * \mathrm{VR}^2) \tag{5.7}$$

These modified values were coded into a new Modelica instance of fluid media. The model was then used to match the inlet conditions for each experimental test with a comparison of the line pressure from the model and experimental result

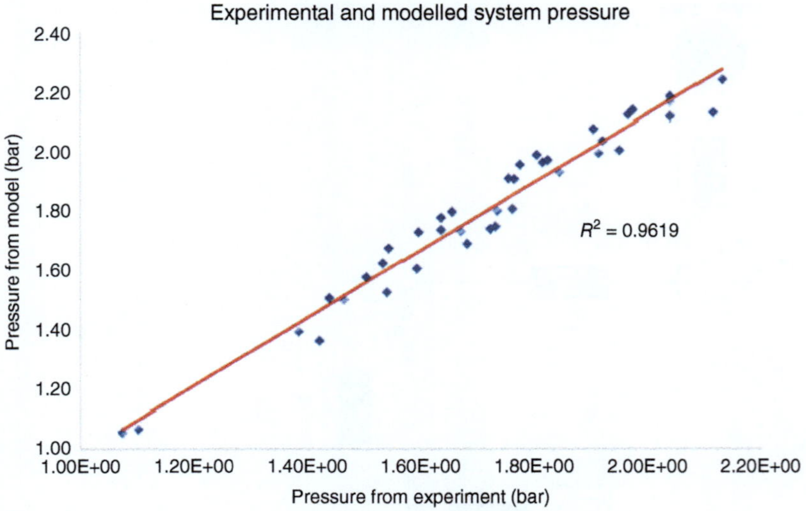

Figure 5.9 Comparison of experimental and modelled line pressure.

shown in Figure 5.9. The percentage difference between the model and experimental result was quantified as:

$$\text{Error} = 2 * \frac{(P_{\text{model}} - P_{\text{exp}})}{(P_{\text{model}} + P_{\text{exp}})} * 100 \quad (5.8)$$

This simple method for estimating the line pressure of this pneumatic conveying system using modified fluid properties was shown to be within 10% of experimental test results across a range of conveying conditions.

5.2.1 Commonly Available Models in Commercial and Open-Source Software

1. Particle trajectory models (One- and two-way coupling)
2. Two fluid models (including Granular Kinetic)
3. Coupled CFD–DEM

An early attempt at modelling pneumatic conveying in a horizontal pipe was presented in Tsuji, Oshima, and Morikawa [8] by tracking the trajectories of individual particles and calculating the fluid drag, the lift force due to particle rotation (Magnus effect), and the torque on the rotation particles. An alternative approach was reported by Mason et al. [9] in their presentation of an algorithm for simulating a complete pneumatic conveying system by a one-dimensional gas–solids flow where the conservation equations were formulated based on the finite volume approach for the gas phase with the effects of the solids given as additional 'source' terms. Levy and Mason [10] also developed a two-layer model for non-suspension gas–solids flows in pipes where the flow in a horizontal pipe was modelled as a dilute gas–solids mixture flowing above a dense gas–solids mixture. The conservation equations for mass and momentum were solved for the gas and solids phases in each layer and the mass and momentum transfers mechanisms between the two layers were modelled.

5.3 Two Fluid Models

The geometry of many pneumatic conveying pipelines with the length-to-diameter ratio in the order of 1000s suggests that there is merit in considering a 1D approach to modelling. One approach is to model the gas and solids as interpenetrating continua as can be found in Mason et al. [11] where the conservation equations are formulated for the gas phase and the effects of the presence of the particulate solids are modelled as source terms in the gas continuity equations. Following their use of a finite volume schematic, Figure 5.10.

Conservation of mass of gas with solids' effect as a source term

$$\dot{m}_{g,e} = \dot{m}_{g,w} + S_r \tag{5.9}$$

$$\rho_{g,e}\alpha_{g,e}A_e u_{g,e} = \rho_{g,w}\alpha_{g,w}A_w u_{g,w} + S_r \tag{5.10}$$

Conservation of momentum of gas with solids' effect as a source term

$$P_e\alpha_{g,e}A_e + \rho_{g,e}\alpha_{g,e}A_e u_{g,e}^2 = P_w\alpha_{g,w}A_w + \rho_{g,w}\alpha_{g,w}A_w u_{g,w}^2 + S_u \tag{5.11}$$

Conservation of energy of gas with solids' effect as a source term

$$\rho_{g,e}\alpha_{g,e}A_e u_{g,e}\left(h_{g,e} + \frac{1}{2}u_{g,e}^2\right) = \rho_{g,w}\alpha_{g,w}A_w u_{g,w}\left(h_{g,w} + \frac{1}{2}u_{g,w}^2\right) + S_h \tag{5.12}$$

Mason et al. [11] used the following for the source terms; no mass transfer between the phases and hence $S_r = 0$, the source term in the momentum equation was based on the determination of forces associated with single particle aerodynamic drag, gravity and wall friction, and the energy source term was determined through enthalpy considerations. No consideration was given to particle–particle interaction. Mason et al. [11] reported reasonable agreement with experimental data in the dilute region when this scheme was implemented in an in-house code.

A 1D model was developed further by Levy et al. [12] to include mass transfer between the phases in an application to the drying of wet solids during transport in a two-phase continuum model. The steady-state flow of wet solid particles was

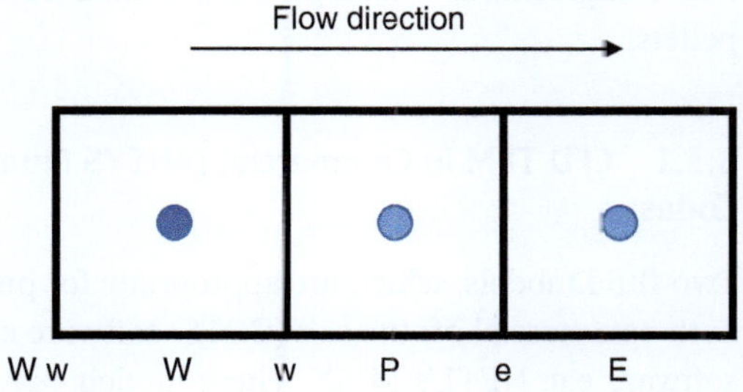

Figure 5.10 Finite volume schematic, centres labelled upper case and faces labelled lower case.

taken as the dispersed phase, and the gas taken as the continuous phase, through the conveying pipeline. The governing equations are obtained by macroscopic mass, momentum, and energy balances for the two phases in a control volume with the following assumptions:

- 'one-dimensional flow assumption can be made,
- mass, momentum and heat transfer occur only between the two phases and not between the particles themselves,
- momentum transfer can occur between the two phases and the pipe wall,
- heat transfer can occur between tile continuous gas phase and the pipe wall,
- the gas phase is a mixture of water vapour with different gases,
- the gas behaves as an ideal gas,
- the particles are made from a substance containing a solid porous matrix and liquid or gas with liquid vapour,
- the densities of the liquid and the solid which compose the particle are constant,
- electrical forces, gravitational forces, and surface tension are neglected.'

The authors demonstrated the experimental results from coal in nitrogen conveying in a high-pressure system were in agreement with the predictions of the numerical simulation.

Levy's 1D model, Levy et al. [12], is also plays a part when combined with CFD–DEM in the prediction of particle attrition in pneumatic conveying systems. Uzi et al. [13].

The one-dimensional approach was extended by Levy [14] to 3D as a 1D model was seen to be insufficient for non-suspension flow by developing a two-fluid model for simulating plug flow in a horizontal pneumatic conveying pipeline. Levy developed his model through continuity equations for each phase and assumed the conveying gas behaved as an ideal gas and the effective normal stress of the solids phase could be described by:

$$P_s = \rho_g R T_g + \sigma_{n0} (\alpha_s / \alpha_{s0})^{1/\beta} \tag{5.13}$$

where σ_{n0} is the value of solids contact stress for a solids volume fraction α_{s0} and β is a constant coefficient over a range of contact pressure. The inter-phase momentum transfer equations included terms for inter-phase friction coefficient that were dependent on local solids concentration and a drag term which took account of multi-particle effects. The equations were solved by the finite volume method in the software package PHEONICS and produced many features of plug flow in agreement with experimental data from conveying tests with polyethylene pellets.

5.3.1 CFD TFM in Commercial (ANSYS Fluent) and Open Source (MFiX) Codes

Two-fluid models, which are appropriate for pneumatic conveying, are available in both commercial off the self (COTS) software e.g. ANSYS Fluent, and open source software e.g. NETL's MFiX. These models may also be referred to as Euler–Euler models where the gas and solids are treated as interpenetrating continua and their respective volume fractions sum to one.

$$\alpha_g + \alpha_s = 1 \tag{5.14}$$

ANSYS recommend either the 'Mixture Model' for dilute (homogeneous) flow or the 'Eulerian Model' dense phase (granular flow) pneumatic conveying and give detailed guidance on selection and use of these models [15]. The two models are summarised in Table 5.1.

5.3.2 Particle Trajectory Models

In these models, the gas phase is treated as a continuum, the Navier–Stokes equations are typically solved using a finite volume method, and solid 'particles' are injected into the system and have their trajectories calculated based on the relevant applied forces, e.g. aerodynamic drag and gravity. This is often cited as an Euler–Lagrange approach. These models are suitable for disperse or dilute phase pneumatic conveying simulations where there is a low volume fraction (also known as low solids loading) and particle-particle interactions can be neglected.

Table 5.1 Comparison of mixture and Eulerian models.

Mixture model	**Eulerian model**
Continuity, momentum and energy equations are solved for the mixture e.g. for mass conservation: $\frac{\partial}{\partial t}(\rho_m) + \nabla \cdot (\rho_m \vec{v}_m) = 0$ (5.15)	Momentum and continuity equations are solved for each phase e.g. for mass conservation: $\frac{\partial}{\partial t}(\alpha_q \rho_q) + \nabla \cdot (\alpha_q \rho_q \vec{v}_q) = \sum_{p=1}^{n} (\dot{m}_{pq} - \dot{m}_{qp}) + S_q$ (5.18)
Where the mass averaged velocity: $\vec{v}_m = \frac{\sum_{k=1}^{n} \alpha_k \rho_k \vec{v}_k}{\rho_m}$ (5.16) And the mixture density: $\rho_m = \sum_{k=1}^{n} \alpha_k \rho_k$ (5.17) For each k phase	The effective density of phase q: $\hat{\rho}_q = \alpha_q \rho_q$ (5.19)
Homogeneous flow (dilute suspension mode flow)	Non-homogeneous flow (dense stratified/plug mode flow)
'Single fluid' approach	Multiple separate, yet interacting phases
Solves for Volume Fraction of the secondary (solids) phase	a scalar transport equation is solved for the secondary-phase volume fraction
Algebraic expressions for relative velocities	Solid-phase shear and bulk viscosities are obtained by applying kinetic theory

If the gas flow field is first calculated and momentum and energy are exchanged with the particles but the corresponding effects of the particles on the gas are not considered, this is known as one-way coupling. This would be applicable to very low solids loading. The other option is where the effects of the particles are taken into account in the calculation of the gas flow is known as two-way coupling.

In these models, the force balance on an individual particle may be written as terms relating to inertia, aerodynamic drag, gravity (buoyancy) and additional source terms (see Section 5.4.1). The ANSYS theory guide [15], for example, writes the force balance

$$\frac{d\vec{u}_p}{dt} = \frac{\vec{u} - \vec{u}_p}{\tau_r} + \frac{\vec{g}(\rho_p - \rho)}{\rho_p} + \vec{F} \tag{5.20}$$

where $\frac{\vec{u}-\vec{u}_p}{\tau_r}$ is the drag force per unit particle mass, and τ_r is the particle relaxation time given by

$$\tau_r = \frac{\rho_p d_p^2}{18\mu} \frac{24}{C_D Re} \tag{5.21}$$

And the relative Reynolds number Re,

$$Re = \frac{\rho d_p |\vec{u}_p - \vec{u}|}{\mu} \tag{5.22}$$

As particle rotation may have a significant effect on the particle's trajectory, an additional ODE for the particle's angular momentum is given as:

$$I_p \frac{d\vec{\omega}_p}{dt} = \frac{\rho}{2}\left(\frac{d_p}{2}\right)^5 C_\omega \vec{\Omega} = \vec{T} \tag{5.23}$$

The relative particle – fluid angular velocity

$$\vec{\Omega} = \frac{1}{2}\nabla \times \vec{u}_f - \vec{\omega}_p \tag{5.24}$$

And assuming a spherical particle, the moment of inertia of the particle:

$$I_p = \frac{\pi}{60}\rho_p d_p^5 \tag{5.25}$$

The additional forces $\vec{F}$ can include a thermophoretic force due to temperature gradients, a Brownian force if modelling sub-micron particles, and Saffman's and Magnus lift forces [15].

5.3.3 Granular Kinetic Theory (GKT)

GKT has been applied to particulate materials and bulk solids; for example, this approach has been used in studying fluidised beds [16] and is appropriate for modelling dense phase pneumatic conveying [3, 17–20].

See Savage and Jeffery [21], Jenkins and Savage [22], Lun et al. [23], Lun [24], and textbooks from Gidaspow [25] and Pannala et al. [26] for background and model development where the kinetic theory of gases is used to describe the behaviour of granular media, for example through the analogue of a thermal temperature (Boltzmann) relation to molecular kinetic energy to a 'granular temperature' related to the kinetic energy of the particles. This approach holds when the particles are in the collision dominant mode. However, unlike the thermal temperature, due to the dissipative nature of granular collisions, the granular media requires an external supply of energy to sustain the inelastic collisions and move to frictional flow or a fixed bed.

This following describes the Eulerian Solids Phase Model with the solids phase represented by granular kinetics formulation as coded by MFiX-two fluid model (MFiX-TFM), (Musser and Carney [27]).

Taking that in this model, the fluid phase volume fraction, ε_f, must be greater than zero, then the sum of solids' volume fractions must always be less than one.

The mass conservation equation for phase m can be expressed as

$$\frac{\partial}{\partial t}(\varepsilon_m \rho_m) + \frac{\partial}{\partial x_j}(\varepsilon_m \rho_m U_{mj}) = \sum_{n=1}^{N_m} R_{mn} + S_m \tag{5.26}$$

where R_{mn} is the rate of production or consumption of the mth phase nth chemical species and users can define a phase mass source term, S_m. In the absence of mass exchange between the phases the RHS = 0 which is the default in MFiX, continuing with default conditions, the momentum conservation equations can be written.

$$\frac{\partial}{\partial t}(\varepsilon_g \rho_g U_{gi}) + \frac{\partial}{\partial x_j}(\varepsilon_g \rho_g U_{gj} U_{gj}) = -\frac{\partial P_g}{\partial x_i} + \frac{\partial \tau_{gij}}{\partial x_j} + \varepsilon_g \rho_g g_i + S_{gi} \tag{5.27}$$

$$\frac{\partial}{\partial t}(\varepsilon_m \rho_m U_{mi}) + \frac{\partial}{\partial x_j}(\varepsilon_m \rho_m U_{mj} U_{mi}) = -\frac{\partial P_m}{\partial x_i} + \frac{\partial \tau_{mij}}{\partial x_j} + \varepsilon_m \rho_m g_i + S_{mi} \tag{5.28}$$

The source terms S_{gi} and S_{mi}, can be recast to differentiate the interphase momentum transfer for other momentum sources.

$$S_{gi} = -\sum_{m=0}^{M} I_{gmi} + \grave{S}_{gi} \tag{5.29}$$

$$S_{mi} = -I_{gmi} + \grave{S}_{mi} \tag{5.30}$$

where the interphase momentum transfer term, I_{gmi}, has the form,

$$I_{gmi} = -\varepsilon_m \frac{\partial P_g}{\partial x_i} + \beta_{gm}(U_{gi} - U_{mi}) \tag{5.31}$$

Or an alternative noted as the Jackson Formulation

$$I_{gmi} = -\varepsilon_m \left(\frac{\partial P_g}{\partial x_i} - \frac{\partial \tau_{gij}}{\partial x_j} \right) + \beta_{gm}(U_{gi} - U_{mi}) \tag{5.32}$$

To simulate pneumatic conveying modes from dense through to dilute, a model must be capable of representing plug or moving bed flow where the particles are in moving or sliding contact analogous to frictional contact, and suspension flow where the particles interact by collision and contacts are short lived. There is no accepted single theory to cover these differing regimes and the transition from one to another. The approach taken in MFiX is that both the frictional and kinetic contributions are calculated and added together based on the 'superposition principle', see also Srivastava and Sundaresan [28].

Frictional regime	**Transition**	**Kinetic regime**
Slow flow		Rapid flow
Strain rate independent		Strain rate dependent
Based on soil mechanics	↔	Based on kinetic theory
Dense phase flow		More dilute phase flow
Contact dominant		Collision dominant

Within MFiX, specific terms relate to the case where the mth phase is a particulate (granular) solid.

The total mth phase solids stress tensor is given by:

$$\tau_{mij} = 2\mu_m D_{mij} + \lambda_m D_{mkk}\delta_{ij} \tag{5.33}$$

where the rate of strain tensor,

$$D_{mij} = \frac{1}{2}\left(\frac{\partial U_{mi}}{\partial x_j} + \frac{\partial U_{mj}}{\partial x_i}\right) \tag{5.34}$$

And the so-called secondary viscosity,

$$\lambda_m = \mu_{\mathrm{b}m} - \frac{2}{3}\mu_m \tag{5.35}$$

And $\mu_{\mathrm{b}m}$ is the bulk viscosity of the mth phase.

Applying the superposition principle of frictional and kinetic components to the solids stress,

$$\tau_{mij} = \tau_{mij}^{\mathrm{fric}} + \tau_{mij}^{\mathrm{kin}} \tag{5.36}$$

Or in terms of viscosity and second viscosity,

$$\mu_m = \mu_m^{\mathrm{fric}} + \mu_m^{\mathrm{kin}} \tag{5.37}$$

$$\lambda_m = \lambda_m^{\mathrm{fric}} + \lambda_m^{\mathrm{kin}} \tag{5.38}$$

And the solids pressure,

$$P_m = P_m^{fric} + P_m^{kin} \tag{5.39}$$

5.3.3.1 The Frictional Regime

Within MFiX, the frictional component only contributes to the solids pressure and solids viscosity when the gas phase volume fraction falls below a set value which represents 'close packed' condition, $\varepsilon_g < \varepsilon^*$., or a 'frictional' condition, $\varepsilon_g < \left(1 - \varepsilon_f^{\mathrm{min}}\right)$.

Syamlal et al. [29] developed a relation for the frictional viscosity for the mth solids phase,

$$\mu_m^{\text{fric}} = \begin{cases} \min\left\{\dfrac{P_c \sin(\phi)}{\sqrt{4I_{m2D}}}\dfrac{\varepsilon_m}{\varepsilon_s}, \mu_s^{\max}\right\} & \varepsilon_g < \varepsilon^* \\ 0 & \varepsilon_g \geq \varepsilon^* \end{cases} \tag{5.40}$$

where;

ϕ is the angle of internal friction,
$\mu_s^{\max}$ is a specified maximum granular viscosity
P_c is the solids pressure
ε_m is the volume fraction of the mth solids phase
ε_s is the total volume fraction of *all* the solids phases
And I_{m2D} is the magnitude of the second invariant of the deviatoric rate of strain tensor.

$$P_c = \begin{cases} A_{pc}(\varepsilon^* - \varepsilon_g)^{n_{pc}} & \varepsilon_g < \varepsilon^* \\ 0 & \varepsilon_g \geq \varepsilon^* \end{cases} \tag{5.41}$$

With default conditions

$n_{pc} = 10$
$A_{pc} = 10^{25}$ Ba (barye)

An alternative model for the frictional regime by Srivastava [28] which

$$\mu_m^{\text{fric}} = \begin{cases} \left\{\dfrac{\sqrt{2}P_f \sin(\phi)}{\sqrt{\dfrac{\Theta_m}{d_m^2} + S_{mij}S_{mji}}}\left(n - (n-1)\left(\dfrac{P_f}{P_c}\right)^{\frac{1}{n-1}}\right)\left(\dfrac{\varepsilon_m}{\varepsilon_s}\right)\right\} & \varepsilon_g < \varepsilon_f^{\min} \\ 0 & \varepsilon_g \geq \varepsilon_f^{\min} \end{cases} \tag{5.42}$$

Additional terms are as follows:
The granular temperature

$$\Theta_m = \frac{1}{3} m_m \left\langle v_{mi}^2 \right\rangle \tag{5.43}$$

d_m is the diameter of particles in the mth phase,
$\varepsilon_f^{\min}$ is the minimum total solids volume required for the onset of frictional stresses.
n is an exponent where for granular assembly dilation equals $\frac{\sqrt{3}}{2\sin(\phi)}$ and for compaction equals 1.03
P_f is the frictional pressure.

5.3.3.2 The Kinetic Regime

In the kinetic regime, individual particles are assumed smooth, frictionless spheres, to move freely and interact by near binary instantaneous collisions which have a constant coefficient of restitution. See Fluent 18.2 Theory Guide or MFiX Theoretical Review for a detailed description and discussion, but example expressions for various terms are given below.

The term p_s, the solids pressure as derived from GKT, is expressed as follows:

$$p_s = \alpha_s \rho_s \Theta_s + 2\rho_s(1 + e_{ss})\alpha_s^2 g_0 \Theta_s, \tag{5.44}$$

where g_0 is a radial distribution function that modifies the probability of particle–particle collisions when the solids phase becomes dense, and, for a single solids phase, it can be written as:

$$g_0 = \left[1 - \left(\frac{\alpha_s}{\alpha_{s,max}}\right)^{1/3}\right]^{-1}, \tag{5.45}$$

and the granular temperature, Θ_m, is proportional to the kinetic energy of the random motion of the particles,

$$\Theta_m = \frac{1}{3} m_m \left\langle v_{mi}^2 \right\rangle \tag{5.46}$$

Based on the findings of van Wachem et al. [30] that the convection and diffusion terms can be neglected from the complete granular temperature transport equation for dense phase flows, and the granular temperature can be solved from the following algebraic equation:

$$0 = (-p_s \bar{\bar{I}} + \bar{\bar{\tau}}_s) : \nabla \vec{v}_s - \gamma_{\Theta_s}, \tag{5.47}$$

where the collisional dissipation of energy, γ_{Θ_s}, is from the correlation by Lun et al. [23]:

$$\gamma_{\Theta_s} = \frac{12\left(1 - e_s^2\right) g_0}{d_s \sqrt{\pi}} \rho_s \alpha_s^2 \Theta_s^{3/2}. \tag{5.48}$$

The solids stress tensor, $\bar{\bar{\tau}}_s$, contains shear and bulk viscosity components plus an additional frictional term to account for the viscous-plastic transition which occurs when the solids volume fraction reaches its maximum value.

$$\mu_s = \mu_{s,col} + \mu_{s,kin} + \mu_{s,fr}. \tag{5.49}$$

The collisional shear viscosity is given by:

$$\mu_{s,col} = \frac{4}{5} \alpha_s \rho_s d_s g_0 (1 + e_s) \left(\frac{\Theta_s}{\pi}\right)^{1/2}. \tag{5.50}$$

and the kinetic viscosity by:

$$\mu_{s,kin} = \frac{10 \rho_s d_s \sqrt{\Theta_s \pi}}{96 \alpha_s (1 + e_s) g_0} \left[1 + \frac{4}{5} g_0 \alpha_s (1 + e_s)\right]^2. \tag{5.51}$$

The bulk viscosity, λ_s, is given as:

$$\lambda_s = \frac{4}{3}\alpha_s^2 \rho_s d_s g_0 (1 + e_s)\sqrt{\frac{\Theta_s}{\pi}} \tag{5.52}$$

5.4 Coupled DEM–CFD Models

5.4.1 Forces on Particles

In a gas–solids flow system, the particle motion and the flow structure are determined by the interactions between gas, particle, and boundary wall. Only part of these interactions are well understood, such as the single particle fluid drag and the normal contact between spherical particles, uncertainties still exist in the scientific understanding of the oblique contact force and multi-particle systems in particular very dense systems. This chapter will start from basic mechanisms of fluid–particle dynamics.

Pneumatic transport of bulk materials has become an important technology in many industries from pharmaceuticals to petro-chemicals and power generation. This is due to the increasing demands for product quality, and of environmental legislation. The bulk pneumatic transport of granular materials in plug mode dense phase has become much more common in recent years. This type of low velocity pneumatic conveying system has been quoted as causing less particle attrition, reduced component wear, and lower energy costs. However, as this technology is being applied more widely to a greater number of materials with higher demands on product quality, a more detailed knowledge of the process is required. This chapter will review this mode of pneumatic conveying.

In bulk solid handling plant and equipment segregation is a common occurrence. Every handling or processing operation gives rise to conditions in which segregation can take place. The consequences of segregation can lead to changes in the materials appearance, texture, taste, density, and so on. Many plant operating problems arise from segregation, such as poor flow, dust, safety hazards, and plant shutdowns. As a consequence, there has been widespread interest in the effect of segregation and mixing on product quality and design of these systems. The mechanisms of segregation have been relatively well understood, but little work has been done on segregation and mixing during dense phase pneumatic conveying in a horizontal pipeline. This chapter reviews mechanisms of segregation and discusses the possibility of occurrence of segregation and mixing in these systems.

This chapter reviews the fundamental mechanisms of fluid-particle dynamics, the conveying mode, dense phase pneumatic conveying, and the mechanisms of segregation. This chapter also identifies the interests and scope of the present study based on this critical review of the current position. Both experimental techniques and mathematical models are discussed in the review and a suitable experimental technique and numerical method are found.

5.4.2 Fluid–Particle Dynamics

In pneumatic conveying, particles are generally either in suspension or dense phase in a turbulent gas stream. The fundamental mechanisms for momentum transfer between gas and particle, which are known as aerodynamic forces have been well established and empirical equations have been correlated for simple engineering problems, such as the fluid drag of a single particle in an infinite fluid or flow through a packed bed. A number of models have been developed for these fundamental mechanisms. This section will discuss these different models and their limitations [31–37].

5.4.3 Single Particle Moving in an Infinite Fluid

5.4.3.1 Fluid-Particle Drag

The force acting on the surface of a particle depends on the flow of the fluid in its immediate vicinity. However, the flow pattern will be influenced by nearby objects such as neighbouring particles. To simplify the problem, it is necessary to consider first a spherical particle flowing at velocity u_r relative to an infinite Newtonian fluid in which there is no other source of shear. It is also assumed that the particle/fluid relative velocity is maintained constant.

As a consequence of the motion of particle in the fluid, a so-called 'drag force' will be exerted on the particle parallel to the particle's relative velocity u_r. Studies of the magnitude of drag force on spheres in steady motion in a fluid date back to Newton's experiments in 1710. Drag force is dependent upon this approach velocity u_r, the fluid density ρ_g the projected area of the particle A_p and drag coefficient C_D

$$F_D = C_D A_p \, {}^1\!/_2 \rho_g u_r^2 \tag{5.53}$$

For a given shape of particle, a simple dimensional analysis would show that C_D is a function of the particle Reynolds number Re_p:

$$Re_p = \frac{\rho_g u_r d_p}{\mu_g} \tag{5.54}$$

where d_p is diameter of the particle and μ_g is viscosity. The particle Reynolds number indicates the nature of flow around the particle. Extensive experiments have been conducted to correlate C_D with Re_p. The relationship between C_D with Re_p can be divided into four regions:

(1) for $Re_p < 20$, $C_D = 24/Re_p$. This is known as Stokes' region. In this low range of Re_p, sometimes called the 'creeping flow' regime, the inertial forces in the basic hydrodynamic Navier–Stokes equations can be neglected
(2) Intermediate region, with $0.25 < Re_p < 500$, C_D is significantly dependent on Re_p,

$$C_D = 18.5 Re_p^{-0.6} \tag{5.55}$$

(3) For $2 \times 10^5 > Re_p > 1000$, $C_D = 0.44$ this is fully developed turbulence regime (Newton's law region);
(4) For $Re_p \geq 2 \times 10^5$, $C_D = 0.1$, this is called turbulent boundary layer region (or supercritical regime).

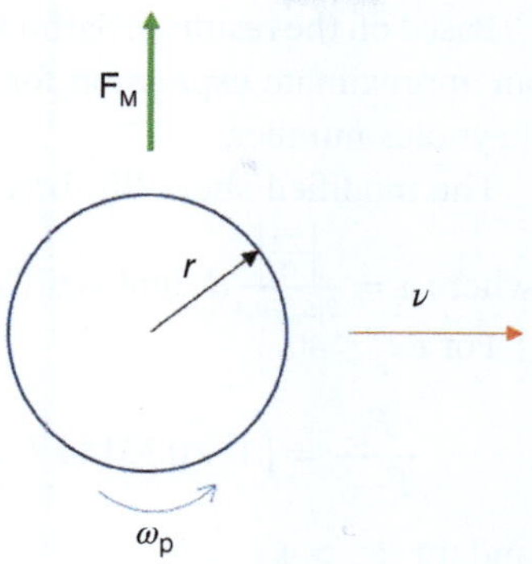

Figure 5.11 Origin of the Magnus force for a ball.

5.4.3.2 Lift Forces

A particle may experience lift forces perpendicular to its translational that are caused by particle rotation and the shear rate in the gas.

Magnus Force The Magnus force [38] is the lift force acting on spinning objects of large size (see Figure 5.11). The lift is caused by a pressure differential between both sides of object resulting from the velocity differential due to rotation.

The lift force is given by,

$$\mathbf{F}_{\mathrm{M}} = \frac{\pi}{8} D_{\mathrm{p}}^{3} \rho \left[\left(\frac{1}{2} \nabla \times \mathbf{u} - \boldsymbol{\omega}_{\mathrm{p}} \right) \times (\mathbf{u} - \mathbf{v}) \right] \tag{5.56}$$

where $\frac{1}{2}\nabla \times \mathbf{u}$ is the local fluid vorticity and $\boldsymbol{\omega}_{\mathrm{p}}$ is the particle rotation. The lift is zero when the particle rotation is equal to the local rotation of the fluid.

The Magnus force acts differently depending on whether the fluid over spinning objects is laminar or turbulent. A turbulent fluid flow results in a lower Magnus force.

The Shear Lift Force In real flow systems, the gas velocity profile is not always uniform at the direction of the main flow. A velocity gradient near the boundary is usually found in any gas flow system. This velocity gradient may produce a shearing force on the particle and result in particle rotation and, at the same time, a transverse force perpendicular to the flow direction is created. Saffman [31] developed a relation for the lift force on a small sphere in a time-independent unbounded linear shear flow, the translation velocity of the particle being parallel to the streamlines and measured relative to the streamline through the centre; this transverse force or lift force F_{L} is described as:

$$F_{\mathrm{L}} = \frac{K\mu(u_{\mathrm{g}} - u_{\mathrm{s}})d^{2}\left|\frac{\mathrm{d}u_{\mathrm{g}}}{\mathrm{d}y}\right|^{\frac{1}{2}} sign\left(\frac{\mathrm{d}u_{\mathrm{g}}}{\mathrm{d}y}\right)}{4(\mu/\rho_{\mathrm{g}})^{\frac{1}{2}}} \tag{5.57}$$

Numerical integration [31] yielded $K = 6.46$.

Saffman [31] assumed that the inertial terms due to shear are large compared to the inertial terms due to the relative velocity and that the particle Reynolds number is small compared to unity.

Based on the results of Saffman [31] and Dandy and Dwyer [32], Mei [33] proposed an approximate expression for the shear lift force on a spherical particle at finite Reynolds number.

The modified shear lift force F_L is valid for $0.005 \leq \alpha < 0.4$ and $0.1 < Re_p < 100$, where $\alpha = \frac{\left|\frac{du_g}{dy}\right|}{2|u_g - u_s|} d_p$ and is divided into two flow regimes as follows:

For $Re_p \leq 40$,

$$\frac{F_L}{F_{L,Sa}} = \left(1 - 0.3314\alpha^{1/2}\right) \exp\left(-\frac{Re_p}{10}\right) + 0.3314\alpha^{1/2} \tag{5.58}$$

and for $Re_p > 40$,

$$\frac{F_L}{F_{L,Sa}} = 0.0524(\alpha Re_p)^{1/2} \tag{5.59}$$

The subscript Sa ($F_{L,Sa}$) denotes the corresponding results obtained by Saffman [31].

5.4.4 Fluid Drag in Multi-Particle System

5.4.4.1 Effect of Voidage/Solid Concentration

The fluid drag force is dependent on not only the relative velocity of fluid and particle but also the presence of other surrounding particles. A simple case of gravitational sedimentation of particles is considered when they are uniformly dispersed in a large volume of gas. It has been observed that the settling velocity is typically much lower than the single-particle terminal velocity. This is due to drag increasing with an increase in particle concentration and the displacement of fluid by the settling particles causing an upflow through the inter-particle voids.

It is extremely difficult to determine this force theoretically. On the other hand, correlations have been well established for the evaluation of this force at either processing equipment or computational cell level.

Ergun [34] studied an extreme situation of the multi-particle system; the solids packed bed, in which the particles remain stationary when a gas flows through. A well-known formula was proposed to predict the pressure gradient which is actually a force per unit length if multiplied by the projected area. This empirical formula fits the experimental data over a wide range of Reynolds numbers. For spherical particles, this takes the form:

$$\frac{\Delta p}{L} = 150\frac{(1-\varepsilon_g)^2}{\varepsilon_g^3}\frac{\mu U_g}{d_s^2} + 1.75\frac{(1-\varepsilon_g)}{\varepsilon_g^3}\frac{\rho_g U_g^2}{d_s} \tag{5.60}$$

Wen-Yu [35] have extended the works of Richardson and Zaki [36] to derive an expression for drag coefficient for mobile particles in fluidization and sedimentation. For voidage of gas greater than 0.8, the drag coefficient is replaced by Wen and Yu's expression.

For individual particles, the drag force acting on a single particle is derived from these correlations.

a. when $\varepsilon_g < 0.8$, using Ergun's equation.

$$\mathbf{f}_i = {}^1\!/\!_2\pi R^2 \rho_g C_D |\mathbf{u} - \mathbf{v}|(\mathbf{u} - \mathbf{v})$$

$$C_D = \varepsilon_g^{-1}\left(\frac{200(1-\varepsilon_g)}{Re} + \frac{7}{3}\right) \tag{5.61}$$

b. when $\varepsilon_g > 0.8$, using Wen and Yu's expression.

$$\mathbf{f}_i = {}^1\!/\!_2\pi R^2 \rho_g C'_D |\mathbf{u} - \mathbf{v}|(\mathbf{u} - \mathbf{v}) \tag{5.62}$$

$$C'_D = C_D \varepsilon_g^{-2.65} \tag{5.63}$$

$$C_D = \begin{cases} \frac{24}{Re}(1 + 0.15Re^{0.687}), & (Re < 1000) \\ 0.44, & (Re \geq 1000) \end{cases} \tag{5.64}$$

$$Re = \frac{\rho_g \varepsilon_g d_p |\mathbf{u} - \mathbf{v}|}{\mu} \tag{5.65}$$

where ρ_g is the fluid density, ε_g is the fluid voidage, μ dynamic fluid viscosity, $\mathbf{v}$ is velocity of particle, d_p the particle diameter, C_D the drag coefficient for an isolated particle, and C'_D the modified drag coefficient,

Di Felice [37] proposed an equation to calculate the fluid drag force acting on a single particle, which is applicable to both fixed and fluidized beds over the full practical range of particle Reynolds number.

$$\mathbf{f}_i = \mathbf{f}_{f0,i}\varepsilon_i^{-\chi} \tag{5.66}$$

where ε_i is the voidage fraction around particle i, taken as the voidage in a computational cell in which particle i is located. The fluid drag force on particle i in the absence of other particle, $\mathbf{f}_{f0,i}$ and the empirical coefficient, χ, are given by

$$\mathbf{f}_{f0,i} = 0.5c_{d0,i}\rho_f \pi R_i^2 |\mathbf{u}_i - \mathbf{v}_i|(\mathbf{u}_i - \mathbf{v}_i) \tag{5.67}$$

and

$$\chi = 3.7 - 0.65\exp\left[-\frac{(1.5 - \log_{10}Re_{p,i})^2}{2}\right] \tag{5.68}$$

where $\mathbf{u}_i$ is the fluid velocity in the computational cell, and $c_{d0,i}$, the fluid drag coefficient, can be expressed as

$$c_{d0,i} = \left(0.63 + \frac{4.8}{Re_{p,i}^{0.5}}\right)^2 \tag{5.69}$$

and $Re_{p,i}$ is the particle Reynolds number

In a pneumatic conveying system, for very dilute phase flow, such that the gas volume fraction is bigger than 0.99, the influence of multi-particle effect is minor, and the fluid drag can be calculated simply using the drag coefficient for a single particle in an infinite fluid. When the solids concentration becomes high, such as in dune flow or plug flow, the Ergun equation, Wen-Yu correlation and Di Felice's equation can be chosen depending on the value of gas voidage and the mode of conveying.

A number of drag coefficients for spheres and their relationship to Re is given in Goossens [39].

5.4.4.2 Non-Spherical Particles

When dealing with non-spherical particles, the term 'particle diameter' or 'particle size' become uncertain and a method must be sought for quantifying the dimensions of the particles and an associated drag coefficient.

A common parameter is the sphericity φ of the particle,

$$\varphi = \frac{s}{S} \tag{5.70}$$

where s is the surface area of the equivalent sphere, and S is the surface area of the actual particle.

This can be used [15] to define the drag coefficient,

$$C_\mathrm{D} = \frac{24}{Re_\mathrm{sph}}\left(1 + b_1 Re_\mathrm{sph}^{b_2}\right) + \frac{b_3 Re_\mathrm{sph}}{b_4 + Re_\mathrm{sph}} \tag{5.71}$$

$$b_1 = \exp(2.3288 - 6.4581\varphi + 2.4486\varphi^2)$$

$$b_2 = 0.0964 + 0.5565\varphi$$

$$b_3 = \exp(4.905 - 13.8944\varphi + 18.4222\varphi^2 - 10.2599\varphi^3)$$

$$b_4 = \exp(1.4681 + 12.2584\varphi - 20.7322\varphi^2 + 15.8855\varphi^3)$$

Another approach given in Dioguardi et al. [40], is a shape factor Ψ,

$$\Psi = \frac{\varphi}{X} \tag{5.72}$$

where X is termed circularity, and is defined by the ratio of the measured particle perimeter to the perimeter of the circle equivalent to the maximum projected area of the particle. Leading to a drag coefficient:

$$C_\mathrm{D} = \frac{0.69 g d_p^3 \rho_\mathrm{f}(1.33\rho_s - 1.33\rho_\mathrm{f})}{\mu_f^2\left(\frac{g\Psi^{1.6} d_p^3 \rho_\mathrm{f}(\rho_s - \rho_\mathrm{f})}{\mu_\mathrm{f}^2}\right)^{1.0412}} \tag{5.73}$$

5.4.5 Particle–Particle Interaction

In laminar flow of very dilute phase flow, particles may not normally collide with each other. However, when the concentration of solids becomes higher and the flow becomes turbulent, particle collisions will inevitably occur.

In a gas–solids system, the significance of particle collisions is important to the flow behaviour and the solids motions. Many researchers have studied this effect using particle contact theory and by analysing the energy loss during the impact. This study will focus on the elastic contact between cohesionless spherical bodies and with the flat boundary wall and neglect surface forces, such as the van der Waals forces, the liquid and solids bridge forces and the electrostatic forces.

5.4.6 Contact Force

5.4.6.1 Normal Contact Force

According to the Hertzian contact theory, the relation between the normal force F_n and displacement δ_n is given by,

$$F_n = K_n \delta_n^{3/2} \tag{5.74}$$

For two different spheres, K_n is given below,

$$K_n = \frac{4\sqrt{r^*}/3}{(1-\sigma_1^2)/E_1 + (1-\sigma_2^2)/E_2}$$

$$r^* = \frac{r_1 r_2}{r_1 + r_2} \tag{5.75}$$

where E is Young's modulus, σ is the Poisson ratio of the particles, and r is the radius of particles.

Walton [41] verified this effect experimentally, theoretically, and numerically with a finite element model and confirmed that both the experiments and the finite calculations are reasonably close to Hertzian contact theory. Walton [41] also discussed inelastic contact in detail. He stated that the elastic–plastic contact initially responds elastically. As the elastic limit is reached, plastic flow occurs and a nearly linear loading behaviour for plastically deforming contacts of spherical surfaces has been observed both experimentally and by numerical model.

5.4.6.2 Tangential Force

Despite more than 100 years having passed since Hertz solved the linear elastic contact equations for the distribution on normal direction stress and strain in the region where two elastic spheres contact each other, Walton [41] stated that there is still no complete solution to the stress and strain distribution if Coulombic friction is assumed to apply on the contact surface, and arbitrary rotation and tangential sliding are combined with changing normal direction contact between elastic spheres. Mindlin [42] and Mindlin and Deresiewice [43] investigated elastic spheres in contact under varying oblique forces. Based on their theories, Tsuji [44] derived a simple formula between tangential force F_t and displacement δ_t for the cases without micro-slip for two spheres,

$$F_t = K_t \delta_t \tag{5.76}$$

$$K_t = \frac{8\sqrt{r^*}\delta_n{}^{1/2}}{(2-\sigma_1)/G_1 + (2-\sigma_2)/G_2}$$

$$r^* = \frac{r_1 r_2}{r_1 + r_2}$$

$$G_1 = \frac{E_1}{2(1+\sigma_1)}$$

$$G_2 = \frac{E_2}{2(1+\sigma_2)} \tag{5.77}$$

It should be noted that if the contact surface is allowed to slip, Tsuji's formula does not adequately describe the tangential force or compliance.

5.5 Summary

The simulation of pneumatic transport of solids (pneumatic conveying) can be undertaken at either system or component level by adopting different models and methods. Three approaches have been described in this chapter: single fluid, two-fluid, and coupled CFD–DEM. The key equations for each approach have been given and illustrated by results from the application of the models to a variety of cases. The capability of both a COTS and an open-source software to simulate pneumatic transport has been discussed.

Model	Typical Use
Single fluid	System pressure drop
1D CFD	Velocity profile
DPM	Dilute phase solid-component interaction, e.g. bend erosion rate estimation
Two-fluid model	Dense phase component level fine particle/powders; moving bed flow behaviour; volume fraction, velocity profile, transient effects.
DEM–CFD	Dense phase component level granular materials; plug/moving bed flow behaviour. Plug formation/collapse; local gas pressure; particle level stress; particle breakage; transient effects; particle shape effects.

Nomenclature

Symbol	Meaning	Dimension (SI units)
ϕ	Solids loading ratio $\frac{\dot{m}_s}{\dot{m}_g}$	—
α	Volume fraction $\frac{V_s}{V_g}$	—
ρ	Density	kg m^{-3}
C_D	Aerodynamic drag coefficient	—
C_ω	Rotational drag coefficient	—
I	Moment of inertia	kg.m^2
λ	Friction factor	—
d	Particle diameter	m
P	Pressure	Pa
U, u, v	Velocity	m s^{-1}
ω	Angular velocity	s^{-1}
ε	Phase volume fraction	—
φ	sphericity	—
Ψ	Shape factor	—

Symbol	Meaning	Dimension (SI units)
SLR	Solids loading ratio	—
CFD	Computational fluid dynamics	
DEM	Discrete element method	
TFM	Two fluid model	
GKT	Granular kinetic theory	
Subscripts		
s	Solids	
p	Particle	
m	Mixture	
f	Fluid	
sph	Sphere	
i,j,m,n	Index values	

References

1 Jones, M.G. and Williams, K.C. (2008). Predicting the mode of flow in pneumatic conveying systems – a review. *Particuology* 6 (5): 289–300.

2 Klinzing, G.E., Marcus, R.D., Rizk, F., and Leung, L.S. (2009). *Pneumatic Conveying of Solids – A Theoretical and Practical Approach.* Springer.

3 Zheng, Y., Pugh, J.R., McGlinchey, D., and Ansell, R.O. (2008). Simulation and experimental study of gas-to-particle heat transfer for non-invasive mass flow measurement. *Measurement* 41 (4).

4 Mills, D. (2004). *Pneumatic Conveying Design Guide*, 2e. Butterworth: Elsevier.

5 Xiang, J., McGlinchey, D., and Latham, J.P. (2010). An investigation of segregation and mixing in dense phase pneumatic conveying. *Granular Matter* 12 (4): 345–355.

6 McGlinchey, D. (2019). Estimation of pneumatic conveying line pressure loss by use of modified fluid properties and modelica simulation. In: *The Proceedings of UK-China International Particle Technology Forum 2019 and UK Annual Particle Technology Forum.* Edinburgh, UK.

7 Folarin, O. (2016). Investigation into the performance and optimisation of pneumatic conveying systems with stepped pipeline. Glasgow Caledoian Univerisity.

8 Tsuji, Y., Oshima, T., and Oshinobu Morikawa, Y. (1985). Numerical simulation of pneumatic conveying in a horizontal pipe. *KONA Powder and Particle Journal* 3: 38–51.

9 Mason, D.J., Marjanovic, P., and Levy, A. (1998). A simulation system for pneumatic conveying systems. In: *Powder Technology*, 7–14. Elsevier Science S.A.

10 Levy, A. and Mason, D.J. (2000). Two-layer model for non-suspension gas–solids flow in pipes. *Powder Technology* 112 (3): 256–262.

11 Mason, D.J., Marjanovic, P., and Levy, A. (1998). POWD A simulation system for pneumatic conveying systems. *Powder Technology* 95.

12 Levy, A., Mason, D.J., Levi-Hevroni, D., and Borde, I. (1998). Drying of wet solid particles in a steady-state one-dimensional flow. *Powder Technology* 95: 15–23.

13 Uzi, A., Kalman, H., and Levy, A. (2016). A novel particle attrition model for conveying systems. *Powder Technology* 298: 30–41.

14 Levy, A. (2000). Two-fluid approach for plug flow simulations in horizontal pneumatic conveying. *Powder Technology* 112.

15 (2019). *ANSYS Fluent User's Guide*. Canonsburg: ANSYS.

16 van Wachem, B.G.M. et al. (1998). Eulerian simulations of bubbling behaviour in gas-solid fluidised beds. *Computers & Chemical Engineering* 22: S299–S306.

17 Ratnayaka, C., Melaaen, M., Datta, B. (2005). Pressure drop prediction in dense phase pneumatic conveying using cfd. In Fourth Int. Conf. on CFD in the Oil and Gas, Metallurgical and Process Industries SINTEF / NTNU Trondheim.

18 McGlinchey, D., Cowell, A., Knight, E.A. et al. (2007). Bend pressure drop predictions using the euler-euler model in dense phase pneumatic conveying. *Particulate Science and Technology* 25 (6): 495–506.

19 Zheng, Y., Pugh, J.R., McGlinchey, D. et al. (2011). *Powder Technology* 208 (1): 231–236.

20 Chun-xia Zhang, Y., Zhang, D.M.G., Yan, D. et al. (2010). Experimental and numerical study on power consumptions in a double-tube-socket pneumatic conveying system. *Powder Technology* 204 (2–3): 268–272.

21 Savage, S.B. and Jeffrey, D.J. (1981). The stress tensor in a granular flow at high shear rates. *Journal of Fluid Mechanics* 110: 255–272.

22 Jenkins, J.T. and Savage, S.B. (1983). A theory for the rapid flow of identical, smooth, nearly elastic, spherical particles. *Journal of Fluid Mechanics* 130: 187–202.

23 Lun, C.K.K., Savage, S.B., Jeffrey, D.J., and Chepurniy, N. (1984). Kinetic theories for granular flow: inelastic particles in Couette flow and slightly inelastic particles in a general flowfield. *Journal of Fluid Mechanics* 140: 223–256.

24 Lun, C.K.K. (1991). Kinetic theory for granular flow of dense , slightly inelastic, slightly rough spheres. *Journal of Fluid Mechanics* 233: 539–559.

25 Gidaspow, D. (1994). *Multiphase Flow and Fluidization: Ontinuum and Kinetic Theory Descriptions*. Academic Press.

26 Pannala, S. et al. (2011). *Computational Gas-Solids Flows and Reacting Systems: Theory, Methods and Practice*. Hershey, PA: Engineering Science Reference.

27 Musser, J. and Carney, J. (2020). Theoretical review of the MFIX fluid and two-fluid models. Morgantown.

28 Srivastava, A. and Sundaresan, S. (2003). Analysis of a frictional-kinetic model for gas-particle flow. *Powder Technology* 129 (1–3): 72–85.

29 Syamlal, M., O'Brien, T.J., and Rogers, W. (1993). *MFIX Documentation Theory Guide*, vol. DOE/METC-9. Morgantown: http://www.osti.gov/servlets/purl/10145548-7h5O7m/native/.

30 van Wachem, B., Schouten, J., van den Bleek, C. et al. (2001). Comparative analysis of CFD models of dense gas solid systems. *AIChE Journal* 47 (5): 1035–1051.

31 Saffman, P.G. (1965). The lift on a small sphere in a slow shear flow. *Journal of Fluid Mechanics* 22 (2): 385–400.

32 Dandy, D.S. and Dwyer, H.A. (1990). A sphere in shear flow at finite reynolds number: effect of shear on particle lift, drag, and heat transfer. *Journal of Fluid Mechanics* 216: 381–410.

33 Mei, R. and Adrian, R. (1992). Flow past a sphere with an oscillation in the free-stream velocity and unsteady drag at finite Reynolds number. *Journal of Fluid Mechanics* 237: 323–341.

34 Ergun, S. (1952). Fluid flow through packed columns. *Chemical Engineering Progress* 48: 9–94.

35 Wen, C.Y. and Yu, Y.H. (1966). Mechanics of fluidization. *Chemical Engineering Progress Symposium Series* 62 (62): 100–111.

36 Richardson, J.F. and Zaki, W.N. (1997). Sedimentation and fluidisation (Trans IChemE, 32:35-53, 1954). *Chemical Engineering Research & Design.*

37 Di Felice, R. (1994). Hydrodynamics of liquid fluidisation. *Chemical Engineering Science* 50 (8): 1213–1245.

38 Lataste, J., Huilier, D., Burnage, H., and Bednář, J. (2000). On the shear lift force acting on heavy particles in a turbulent boundary layer. *Atmospheric Environment* 34 (23): 3963–3971.

39 Goossens, W.R.A. (2019). Review of the empirical correlations for the drag coefficient of rigid spheres. *Powder Technology*. Elsevier B.V. 352: 350–359.

40 Dioguardi, F., Dellino, P., and Mele, D. (2014). Integration of a new shape-dependent particle-fluid drag coefficient law in the multiphase Eulerian-Lagrangian code MFIX-DEM. *Powder Technology* 260: 68–77.

41 Walton, O.R. (1993). Numerical Simulation of Inclined Chute Flows of Monodisperse, Inelastic, Frictional Spheres. *Mechanics of Materials* 16 (1): 239–247.

42 Mindlin, R.D. (1949). Compliance of elastic bodies in contact. *Journal of Applied Mechanics* 16: 259–268.

43 Mindlin, R.D. and Deresiewicz, H. (1953). Elastic spheres in contact under varying oblique forces. *Transactions of ASME, Series E, Journal of Applied Mechanics* 20: 327–344.

44 Tsuji Y., Tanaka T., Ishida T. Lagrangian numerical simulation of plug flow of cohesionless particles in a horizontal pipe (1992) Powder Technology, 71 (3), pp. 239–250 DOI: 10.1016/0032-5910(92)88030-L

Further Reading

Ariyaratne, W.K.H., Ratnayake, C., and Melaaen, M.C. (2019). CFD modeling of dilute phase pneumatic conveying in a horizontal pipe using Euler–Euler approach. *Particulate Science and Technology* 37 (8): 1011–1019.

Babaoğlu, N.U., Parvaz, F., Hosseini, S.H. et al. (2021). Influence of the inlet cross-sectional shape on the performance of a multi-inlet gas cyclone. *Powder Technology* 384: 82–99.

Balzer, G. (2000). Gas-solid flow modelling based on the kinetic theory of granular media: validation, applications and limitations. *Powder Technology* 113: 299–309.

Benyahia, S. (2008). A time-averaged model for gas-solids flow in a one-dimensional vertical channel. *Chemical Engineering Science* 63 (9): 2536–2547.

Chen, X.Z., Shi, D.P., Gao, X., and Luo, Z.H. (2011). A fundamental CFD study of the gas-solid flow field in fluidized bed polymerization reactors. *Powder Technology* 205 (1–3): 276–288.

Curtis, J.S. and Van Wachem, B. (2004). Modeling particle-laden flows: a research outlook. *AIChE Journal* 50: 2638–2645.

da Silva Rodrigues, S. (2014). Riser hydrodynamic study with group B particles for chemical looping combustion. *Chemical Engineering* .

Dartevelle S. (2003). Numerical and granulometric approaches to geophysical granular flows. Michigan Technological University.

De Wilde, J., Van Engelandt, G., Heynderickx, G.J., and Marin, G.B. (2005). Gas-solids mixing in the inlet zone of a dilute circulating fluidized bed. *Powder Technology* 151: 96–116.

Du, W., Bao, X., Xu, J., and Wei, W. (2006). Computational fluid dynamics (CFD) modeling of spouted bed: assessment of drag coefficient correlations. *Chemical Engineering Science* 61 (5): 1401–1420.

Duarte, C.A.R. and de Souza, F.J. (2017). Innovative pipe wall design to mitigate elbow erosion: a CFD analysis. *Wear* 380–381: 176–190.

Duarte, C.A.R., de Souza, F.J., Salvo, R. de, V., and dos Santos, V.F. (2017). The role of inter-particle collisions on elbow erosion. *International Journal of Multiphase Flow* 89: 1–22.

Ebrahimi, M. and Crapper, M. (2017). CFD–DEM simulation of turbulence modulation in horizontal pneumatic conveying. *Particuology* 31: 15–24.

El-Askary, W.A., Eldesoky, I.M., Saleh, O. et al. (2016). Behavior of downward turbulent gas-solid flow through sudden expansion pipe. *Powder Technology* 291: 351–365.

El-Behery, S.M., El-Askary, W.A., Hamed, M.H., and Ibrahim, K.A. (2013). Eulerian-Lagrangian simulation and experimental validation of pneumatic conveying dryer. *Drying Technology* 31 (12): 1374–1387.

El-Emam, M.A., Zhou, L., Shi, W., and Han, C. (2021). Performance evaluation of standard cyclone separators by using CFD–DEM simulation with realistic bio-particulate matter. *Powder Technology* 385: 357–374.

Evrard, F., Denner, F., and van Wachem, B. (2020). Euler-Lagrange modelling of dilute particle-laden flows with arbitrary particle-size to mesh-spacing ratio. *Journal of Computational Physics: X* 8: 100078.

Fatahian, E., Fatahian, H., Hosseini, E., and Ahmadi, G. (2021). A low-cost solution for the collection of fine particles in square cyclone: a numerical analysis. *Powder Technology* 387: 454–465.

Foroozesh, J., Parvaz, F., Hosseini, S.H. et al. (2021). Computational fluid dynamics study of the impact of surface roughness on cyclone performance and erosion. *Powder Technology* 389: 339–354.

Fritzson, P., Pop, A., Abdelhak, K. et al. (2020). The OpenModelica integrated environment for modeling, simulation, and model-based development. *Modeling Identification and Control* 41 (4): 241–285.

Fritzson et al. (2020). *Modeling, Identification and Control* 41 (4): 241–285.

Fu, S., Zhou, F., Sun, G. et al. (2021). Performance evaluation of industrial large-scale cyclone separator with novel vortex finder. *Advanced Powder Technology* 32 (3): 931–939.

Gidaspow, D., Jung, J., and Singh, R.K. (2004). Hydrodynamics of fluidization using kinetic theory: an emerging paradigm 2002 Flour-Daniel lecture. *Powder Technology* 148: 123–141.

Goldschmidt, M.J.V., Beetstra, R., and Kuipers, J.A.M. (2002). Hydrodynamic modelling of dense gas-fluidised beds: comparison of the kinetic theory of granular row with 3D hard-sphere discrete particle simulations. *Chemical Engineering Science* 57: 2059–2075.

Gundogdu, M.Y., Kutlar, A.I., and Duz, H. (2009 Jan). Analytical prediction of pressure loss through a sudden-expansion in two-phase pneumatic conveying lines. *Advanced Powder Technology* 20 (1): 48–54.

Henthorn, K.H., Park, K., and Curtis, J.S. (2005). Measurement and prediction of pressure drop in pneumatic conveying: effect of particle characteristics, mass loading, and Reynolds number. *Industrial and Engineering Chemistry Research* 44 (14): 5090–5098.

Hilton, J.E. and Cleary, P.W. (2011). The influence of particle shape on flow modes in pneumatic conveying. *Chemical Engineering Science* 66 (3): 231–240.

Hong, J. and Tomitat, Y. (1995). Analysis of high density gas-solids stratified pipe flow. *International Journal of Multiphase Flow* 21.

Hua, L., Zhao, H., Li, J. et al. (2015). Eulerian-Eulerian simulation of irregular particles in dense gas-solid fluidized beds. *Powder Technology* 284: 299–311.

Huanpeng, L., Wentie, L., Jianxiang, Z. et al. (2004). Numerical study of gas-solid flow in a precalciner using kinetic theory of granular flow. *Chemical Engineering Journal* 102 (2): 151–160.

Huilin, L., Gidaspow, D., Bouillard, J., and Wentie, L. (2003). Hydrodynamic simulation of gas-solid flow in a riser using kinetic theory of granular flow. *Chemical Engineering Journal* 95 (1–3): 1–13.

Ji, Y., Liu, S., and Li, J. (2018). Experimental and numerical studies on dense-phase pneumatic conveying of spraying material in venturi. *Powder Technology* 339: 419–433.

Kalman, H., Portnikov, D., Gabrieli, O.G., and Tripathi, N.M. (2019). What do pneumatic conveying and hydraulic conveying have in common? *Powder Technology* 354: 485–495.

Kalman, H. and Rawat, A. (2020). Flow regime chart for pneumatic conveying. *Chemical Engineering Science* 211.

Kang, R. and Liu, H. (2020). An integrated model of predicting sand erosion in elbows for multiphase flows. *Powder Technology* 366: 508–519.

Kapusuzoglu, B. and Mahadevan, S. (2021). Information fusion and machine learning for sensitivity analysis using physics knowledge and experimental data. *Reliability Engineering and System Safety* 214: 107712.

Kartushinsky, A., Tisler, S., Oliveira, J.L.G., and van der Geld, C.W.M. (2016). Eulerian-Eulerian modelling of particle-laden two-phase flow. *Powder Technology* 301: 999–1007.

Kaushal, D.R., Thinglas, T., Tomita, Y. et al. (2012). CFD modeling for pipeline flow of fine particles at high concentration. *International Journal of Multiphase Flow* 43: 85–100.

Kim, Y.N., Wu, C., and Cheng, Y. (2011). CFD simulation of hydrodynamics of gas-solid multiphase flow in downer reactors: revisited. *Chemical Engineering Science* 66 (21): 5357–5365.

Klinzing, G.E. (2018). Historical review of pneumatic conveying. *KONA Powder and Particle Journal.* . Hosokawa Powder Technology Foundation 2018: 150–159.

Klinzing, G.E. (2018). A review of pneumatic conveying status, advances and projections. *Powder Technology*. Elsevier B.V. 333: 78–90.

Kozołub, P., Klimanek, A., Białecki, R.A., and Adamczyk, W.P. (2017). Numerical simulation of a dense solid particle flow inside a cyclone separator using the hybrid Euler–Lagrange approach. *Particuology* 31: 170–180.

Kuang, S., Zhou, M., and Yu, A. (2020). CFD-DEM modelling and simulation of pneumatic conveying: a review. *Powder Technology* 365: 186–207.

Kuang, S.B., Li, K., Zou, R.P. et al. (2013). Application of periodic boundary conditions to CFD-DEM simulation of gas-solid flow in pneumatic conveying. *Chemical Engineering Science* 93: 214–228.

Kuang, S.B., Zou, R.P., Pan, R.H., and Yu, A.B. (2012). Gas-solid flow and energy dissipation in inclined pneumatic conveying. *Industrial and Engineering Chemistry Research* 51 (43): 14289–14302.

Laı´ n, S., Sommerfeld, M., and Kussin, J. Experimental studies and modelling of four-way coupling in particle-laden horizontal channel flow. *International Journal of Heat and Fluid Flow* 23 (5): 647–656.

Lan, X., Xu, C., Gao, J., and Al-Dahhan, M. (2012). Influence of solid-phase wall boundary condition on CFD simulation of spouted beds. *Chemical Engineering Science* 69 (1): 419–430.

Lavrinec, A., Orozovic, O., Rajabnia, H. et al. (2020). Velocity and porosity relationships within dense phase pneumatic conveying as studied using coupled CFD-DEM. *Powder Technology* 375: 89–100.

Lim, E.W.C., Wang, C.H., and Yu, A.B. (2006). Discrete element simulation for pneumatic conveying of granular material. *AIChE Journal* 52 (2): 496–509.

Lin, Z., Sun, X., Yu, T. et al. (2020). Gas–solid two-phase flow and erosion calculation of gate valve based on the CFD-DEM model. *Powder Technology* 366: 395–407.

Lungu, M., Siame, J., and Mukosha, L. (2021). Comparison of CFD-DEM and TFM approaches for the simulation of the small scale challenge problem 1. *Powder Technology* 378: 85–103.

Mallouppas, G. and van Wachem, B. (2013). Large Eddy simulations of turbulent particle-laden channel flow. *International Journal of Multiphase Flow* 54: 65–75.

Manjula, E.V.P.J., Ariyaratne, W.K.H., Ratnayake, C., and Melaaen, M.C. (2017). A review of CFD modelling studies on pneumatic conveying and challenges in modelling offshore drill cuttings transport. *Powder Technology*. Elsevier B.V. 305: 782–793.

Marcus, R.D., Leung, L.S., Klinzing, G.E., and Rizk, F. (1993). Pneumatic conveying of solids: a theoretical and practical approach. *Drying Technology* 11: 859–860.

Marjanovic, P., Levy, A., and Mason, D.J. (1999). An investigation of the flow structure through abrupt enlargement of circular pipe. *Powder Technology* 104: 296–303.

McGlinchey, D., Cowell, A., and Crowe, R. (2012). CFD investigation of dense phase pneumatic conveying at a pipeline enlargement. *Particuology* 10 (2): 176–183.

Mezhericher, M., Brosh, T., and Levy, A. (2011). Modeling of particle pneumatic conveying using DEM and DPM methods. *Particulate Science and Technology* 29 (2): 197–208.

Miao, Z., Kuang, S., Zughbi, H., and Yu, A. (2019). CFD simulation of dilute-phase pneumatic conveying of powders. *Powder Technology* 349: 70–83.

Nikku, M., Jalali, P., Ritvanen, J., and Hyppänen, T. (2014). Characterization method of average gas-solid drag for regular and irregular particle groups. *Powder Technology* 253: 284–294.

Panday, R., Shadle, L.J., Shahnam, M. et al. (2014). Challenge problem: 1. Model validation of circulating fluidized beds. *Powder Technology* 258: 370–391.

Peng, W., Cao, X., Hou, J. et al. (2020). Experiment and numerical simulation of sand particle erosion under slug flow condition in a horizontal pipe bend. *Journal of Natural Gas Science and Engineering* 76: 103175.

Rajan, K.S., Srivastava, S.N., Pitchumani, B., and Mohanty, B. (2006). Simulation of gas-solid heat transfer during pneumatic conveying: use of multiple gas inlets along the duct. *International Communications in Heat and Mass Transfer* 33 (10): 1234–1242.

Rong, L.W., Zhou, Z.Y., and Yu, A.B. (2015). Lattice-Boltzmann simulation of fluid flow through packed beds of uniform ellipsoids. *Powder Technology* 285: 146–156.

Safikhani, H., Akhavan-Behabadi, M.A., Shams, M., and Rahimyan, M.H. (2010). Numerical simulation of flow field in three types of standard cyclone separators. *Advanced Powder Technology* 21 (4): 435–442.

Sakai, M. and Koshizuka, S. (2009). Large-scale discrete element modeling in pneumatic conveying. *Chemical Engineering Science* 64 (3): 533–539.

Sansana, J., Joswiak, M.N., Castillo, I. et al. (2021). Recent trends on hybrid modeling for industry 4.0. *Computers and Chemical Engineering* (151): 107365.

Snider, D.M. (2001). An incompressible three-dimensional multiphase particle-in-cell model for dense particle flows. *Journal of Computational Physics* 170 (2): 523–549.

Sommerfeld, M. Validation of a stochastic Lagrangian modelling approach for inter-particle collisions in homogeneous isotropic turbulence. *International Journal of Multiphase Flow*, 27 (10): 1829–1858.

Sommerfeld, M. and Kussin, J. (2003). Analysis of collision effects for turbulent gas-particle flow in a horizontal channel. Part II. Integral properties and validation. *International Journal of Multiphase Flow* 29 (4): 701–718.

Sommerfeld, M. and Lain, S. (2015). Parameters influencing dilute-phase pneumatic conveying through pipe systems: a computational study by the Euler/Lagrange approach. *Canadian Journal of Chemical Engineering* 93 (1): 1–17.

Sun, S., Yuan, Z., Peng, Z. et al. (2018). Computational investigation of particle flow characteristics in pressurised dense phase pneumatic conveying systems. *Powder Technology* 329: 241–251.

Syamlal M, O'brien TJ. (n.d.). The derivation of a drag coefficient formula from velocity-voidage correlations. Unpublished.

Syamlal, M., Rogers, W.A., and O'brien, T.J. (1993). MFIX Documentation: Theory Guide. U.S. Department of Energy.

Tardos, G.I. (1997). A fluid mechanistic approach to slow, frictional flow of powders. *Powder Technology* 92 (1): 61–74.

Uzi, A., Ben Ami, Y., and Levy, A. (2017). Erosion prediction of industrial conveying pipelines. *Powder Technology* 309: 49–60.

van Wachem, B., Zastawny, M., Zhao, F., and Mallouppas, G. (2015). Modelling of gas-solid turbulent channel flow with non-spherical particles with large Stokes numbers. *International Journal of Multiphase Flow* 68: 80–92.

Vollmari, K., Jasevičius, R., and Kruggel-Emden, H. (2016). Experimental and numerical study of fluidization and pressure drop of spherical and non-spherical particles in a model scale fluidized bed. *Powder Technology* 291: 506–521.

Wang, Y., Williams, K., Jones, M., and Chen, B. (2017). CFD simulation methodology for gas-solid flow in bypass pneumatic conveying – A review. *Applied Thermal Engineering* 125: 185–208.

Wei, Z.J., Liu, Y., Yong, L.S. et al. (2017). Effects of particle shape and swirling intensity on elbow erosion in dilute-phase pneumatic conveying. *Wear* 380–381: 66–77.

Xu, L., Zhang, Q., Zheng, J., and Zhao, Y. (2016). Numerical prediction of erosion in elbow based on CFD-DEM simulation. *Powder Technology* 302: 236–246.

Yang, D., Xing, B., Li, J. et al. (2019). Experiment and simulation analysis of the suspension behavior of large (5–30 mm) nonspherical particles in vertical pneumatic conveying. *Powder Technology* 354: 442–455.

Zamani, M., Seddighi, S., and Nazif, H.R. (2017). Erosion of natural gas elbows due to rotating particles in turbulent gas-solid flow. *Journal of Natural Gas Science and Engineering* 40: 91–113.

Zhang, Y. and Reese, J.M. (2001). Particle±gas turbulence interactions in a kinetic theory approach to granular flows. *International Journal of Multiphase Flow* 27 (11): 1945–1964.

Zhang Y, Reese JM. The influence of the drag force due to the interstitial gas on granular flows down a chute.

Zhang, Z., Dong, S., Dong, K. et al. (2022). Experimental and numerical study of a gas cyclone with a central filter. *Particuology* 63: 47–59.

Zhao, H. and Zhao, Y. (2019). CFD–DEM simulation of pneumatic conveying in a horizontal channel. *International Journal of Multiphase Flow* 118: 64–74.

Zhou, F., Hu, S., Liu, Y. et al. (2014). CFD-DEM simulation of the pneumatic conveying of fine particles through a horizontal slit. *Particuology* 16: 196–205.

Zhu, H.P., Zhou, Z.Y., Yang, R.Y., and Yu, A.B. (2008). Discrete particle simulation of particulate systems: a review of major applications and findings. *Chemical Engineering Science* 63: 5728–5770.

6

Modelling and Simulation of Erosive Wear and Particle Breakage in Pneumatic Conveying

Rahul Tarodiya and Avi Levy

Ben-Gurion University of the Negev, Department of Mechanical Engineering, PO Box 653, Beer-Sheva 81405, Israel

6.1 Introduction

Pneumatic conveying of solids has been commonly accepted as an effective and cost-saving method of transporting bulk solids. However, the operation involves a significant amount of particle-to-particle and particle-to-wall collisions. As a result, energy is transferred to the impact location, which may initiate the erosion of the pipeline or cause particle breakage. These are widely seen as an unavoidable phenomenon influencing the operating efficiencies of the pneumatic system. Furthermore, particle breakage affects the product quality, whereas erosion leads to frequent maintenance and replacement of equipment. Thus, the phenomenon of erosive wear and particle breakage, especially at high velocity, must often be assessed in conjunction with system design or operation. In addition to experimental investigations, numerical modelling can also be used to properly capture these phenomena. In fact, numerical modelling for erosive wear and particle breakage is becoming more common in the designing of the pneumatic conveying system.

This chapter aims to discuss different methodologies that can adopt to computationally capture particle breakage and erosive wear in pneumatic conveying. In subsequent sections, one-dimensional, two-phase flow modelling of the industrial conveying pipeline for particle breakage and erosion prediction is presented in detail, including the algorithm for breakage and erosion modelling and application examples employing the algorithms to simulate the conveying line. It is one of the effective approaches to simulate the particle breakage and erosive wear of a large-scale industrial conveying line at less computational time.

6.1.1 Erosive Wear

Erosion due to solid particle impact is defined as the progressive mass loss, fracture, or displacement of material due to repetitive impact of particles on the target

Simulations in Bulk Solids Handling: Applications of DEM and other Methods, First Edition.
Edited by Don McGlinchey.

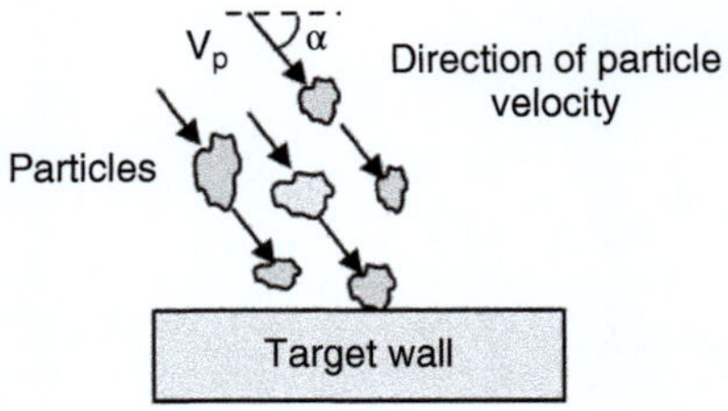

Figure 6.1 Solid particles impact on target wall.

surface. It involves kinetic energy transformation of the impacting particle to the target wall, as schematically shown in Figure 6.1. It is a complex phenomenon affected by the individual or combined effect of change in process parameters such as particle properties, target surface properties, and particle impact conditions.

In pneumatic conveying of solids, the component failure due to erosive wear is mainly classified into two modes: (i) sliding-abrasion and (ii) particle-impact. Sliding-abrasion is a predominant mode of material removal in a dense particulate flow. It involves a bed of contact-load particle bearing against a surface and moving tangent to it, as illustrated in Figure 6.2a. Particle-impact involves the collision of individual particles on the target surface at an angle due to turbulence. It can be further categorised into random or directional impact, as illustrated in Figure 6.2b,c. Depending upon the flow conditions, the dominance of these impact conditions varies.

6.1.2 Particle Breakage

Particle breakage is defined as the separation of a single (mother) particle into two or more particles (sibling). Based on the industrial need, particle breakage can be either desirable (comminution) or undesirable (attrition). It occurs in a variety of modes depending upon the stress conditions and the material properties. In general, it can be categorised based on two breakage mechanisms i.e. abrasion/chipping and fragmentation, which are related to the applied stresses. In practice, fragmentation is associated with the large-scale particle fracture that may occur when the applied stresses acting on the particle exceed some critical value. It results in wide size distribution, and the mother particle loses its identity. Abrasion/chipping is associated with the small-scale damage at relatively small applied stresses. Abrasion follows the forces acting parallel to the surface of the particles, whereas chipping is due to oblique forces. In both, a very small amount of material is removed from the particle surface where the critical stress exceeds locally. The mother particle retains its

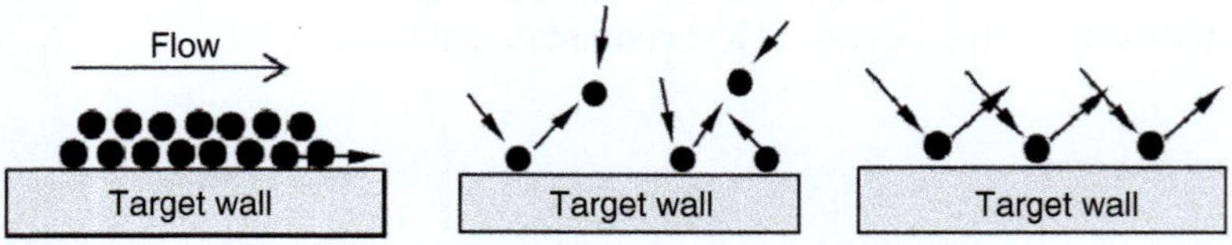

Figure 6.2 Modes of particle impact on component wall producing erosion. (a) Sliding bed. (b) Random impact. (c) Directional impact.

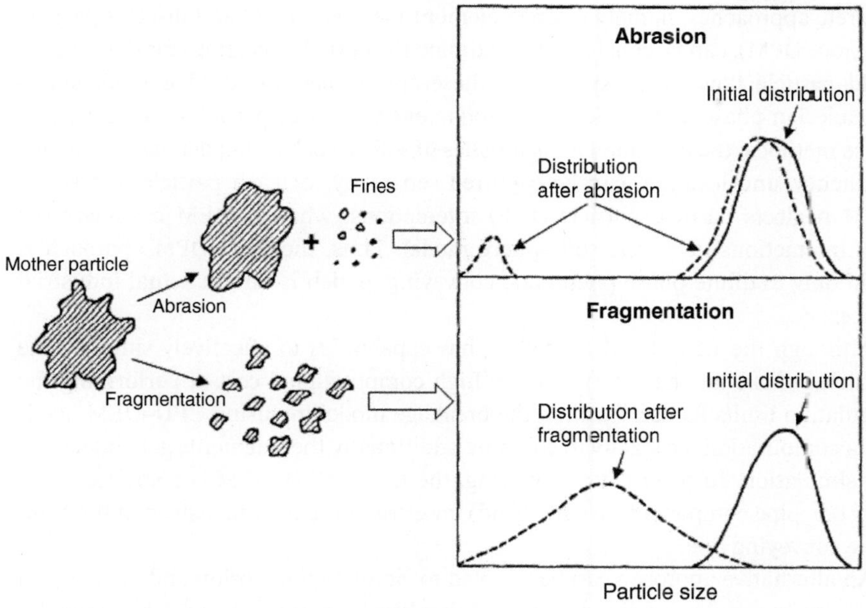

Figure 6.3 Schematic definition of particle breakage.

identity but experiences a slow and progressive loss of mass. Figure 6.3 illustrates the particle breakage associated with abrasion and fragmentation.

6.1.3 Modelling Approaches to Simulate Pneumatic Conveying Pipeline

With the continuous progress in computational technology over the last two decades, numerical modelling has emerged as a key method for tackling the issues of particle breakage and surface erosion. It provides an alternative to expensive and laborious large-scale experiments. Additionally, it allows for an independent examination of the influencing parameters on the process, which is difficult to achieve with experiments. Furthermore, experiments cannot extract the same level of details that simulations can. For example, the common way to experimentally analyse the particle breakage for a specific system is by comparing the particle size distribution (PSD) at the entrance to that at the outflow. However, correlating this information to the system operation is extremely difficult because of the continuous change in PSD along the flow path in the pipeline, which is regulated by a large number of parameters. Moreover, despite the advantages of numerical modelling, to obtain detailed information or system design and optimization, experimental measurements are the only appropriate means to inspect the reliability of the numerical models.

The modelling of particle breakage and surface erosion for pneumatic conveying can be performed with the estimation of particle dynamics in the system and their incorporation in the breakage and erosion prediction model. CFD-coupled discrete approaches, namely discrete element method (DEM) and discrete particle method (DPM), can be considered to estimate the particle dynamics for the coupled fluid–particle flow in the system. In these approaches, the fluid is modelled as an Eulerian phase and particles as discrete entities. Since particles are discrete in these methods, the collision characteristics of solids, such as impact angle, velocity, frequency, and location, can be captured separately for each particle. Moreover, DPM neglects particle–particle (P–P) interactions, whereas DEM considers the P–P interactions using the soft sphere model. Thus, the CFD–DPM approach is valid only to dilute phase pneumatic conveying, which is rare in actual industrial processes.

Although the CFD–DEM modelling has capabilities to effectively simulate the particle collisional characteristics, the high computational cost of performing the simulation limits its use. Further, the breakage modelling using CFD–DEM needs extra computational capacity to compute additionally the fragments generated during simulation. In pneumatic conveying, the use of CFD–DEM is restricted up to only one pipe component (usually bend) for erosion prediction, instead of the complete conveying line.

An alternative approach can be adopted to simulate the erosion and breakage in the complete conveying line by using a simplified macroscopic model. It includes modeling of two-phase flow by using a zero- or one-dimensional approach and incorporation of empirical models for erosion and breakage prediction. In zero-dimensional flow modelling [1], the travelling velocity of the particles is considered constant. Thus, the particle impact velocity is always the same. In one-dimensional modelling, the travelling velocity of the particles along the flow direction is obtained by solving a conventional one-dimensional, two-phase flow model. Additionally, the machine functions are utilized to account for collision characteristics. These are the statistical functions that depend on the geometry and operating conditions. These can be obtained from the particle collision data collected from the CFD–DEM simulations.

Further, for erosion predictions, proper selection of a model can be done from several available models based on the material properties and operating conditions. For breakage prediction, three approaches are generally considered: (i) the use of empirical comminution functions [2, 3], (ii) the use of an empirical breakage model [4], and (iii) the use of a bonded particle model [5].

6.2 One-dimensional Particle Breakage Modelling of the Conveying Pipeline

The one-dimensional, two-phase flow modelling together with the breakage algorithm and functions for particle collision characteristics and comminution is a relatively quick approach to simulate the conveying pipeline for particle breakage assessment. For calculating the two-phase flow dynamics, a conventional one-dimensional, two-phase flow model is used that resolves solid- and gas-phase velocities, pressure, temperature, and volume fraction. Second, the one-dimensional breakage algorithm (ODBA) is employed to account for breakage. Subsequent sections discuss the details regarding the one-dimensional, two-phase flow model and the ODBA.

6.2.1 One-dimensional Two-phase Flow Model

The one-dimensional modelling of the two-phase flow is initiated with the division of pipeline geometry into numerical cells of equal size. The model is then solved to obtain two-phase flow dynamics under the following assumptions: (i) Flow is steady, (ii) Air is an ideal gas, (iii) The solid holdup of the particles is equivalent to the particles' cross-sectional coverage area, and (iv) The particles are homogenously suspended in the pipe cross-section.

The two-phase flow dynamics for each differential segment along the pipeline are determined by following the conservation laws of mass, momentum, and energy.

The mass balance equations for gas and particle phases are given as:

$$\frac{\varepsilon_g v_g \rho_g A}{dx} = 0 \tag{6.1}$$

$$\frac{\varepsilon_p v_p \rho_p A}{dx} = 0 \tag{6.2}$$

where

ρ_g = gas phase density
ρ_p = particle phase density
v_g = gas phase velocity
v_p = particle phase velocity
A = pipe cross-sectional area

The volumetric fractions are defined as ε_g and ε_p of gas and particle phases, respectively, the sums of which must be unity, i.e. $\varepsilon_g + \varepsilon_p = 1$.

The momentum balance equations for gas and particle phases are given as:

$$\frac{d\left(\varepsilon_g v_g^2 \rho_g A\right)}{dx} = \frac{dP}{dx}A - \tau_{wg}S - \varepsilon_g \rho_g A(g \sin\theta) + \beta(v_p - v_g) \tag{6.3}$$

$$\frac{d\left(\varepsilon_p v_p^2 \rho_p A\right)}{dx} = -\tau_{wp}S - \varepsilon_p \rho_p A(g \sin\theta) - \beta(v_p - v_g) \tag{6.4}$$

where

S = pipe circumference
P = operative local static pressure = $\rho_g \mathfrak{R} T_g$, where $\mathfrak{R}$ is the ideal gas constant and T_g is the gas temperature
g = gravitational acceleration
τ_{wg} = friction of the gas phase with the wall
τ_{wp} = particle friction with the wall
β = interface friction coefficient

The energy balance equations for the gas and particle phase are given as:

$$\frac{\mathrm{d}\left(C_{\mathrm{p,g}}T_{\mathrm{g}}\varepsilon_{\mathrm{g}}\rho_{\mathrm{g}}v_{\mathrm{g}}A+1/2\varepsilon_{\mathrm{g}}v_{\mathrm{g}}^{3}\rho_{\mathrm{g}}A\right)}{\mathrm{d}x}=\beta(v_{\mathrm{p}}-v_{\mathrm{g}})v_{\mathrm{p}}-\varepsilon_{\mathrm{g}}\rho_{\mathrm{g}}A(g\sin\theta)v_{\mathrm{g}}-h_{\mathrm{pg}}N_{\mathrm{p}}\pi\mathrm{d}^{2}(T_{\mathrm{g}}-T_{\mathrm{p}})A+h_{\mathrm{wg}}(T_{\mathrm{w}}-T_{\mathrm{g}}) \tag{6.5}$$

$$\frac{\mathrm{d}\left(C_{\mathrm{p,p}}T_{\mathrm{p}}\varepsilon_{\mathrm{p}}\rho_{\mathrm{p}}v_{\mathrm{p}}A+1/2\varepsilon_{\mathrm{p}}v_{\mathrm{p}}^{3}\rho_{\mathrm{p}}A\right)}{\mathrm{d}x}=-\beta(v_{\mathrm{p}}-v_{\mathrm{g}})v_{\mathrm{p}}-\varepsilon_{\mathrm{p}}\rho_{\mathrm{p}}A(g\sin\theta)v_{\mathrm{p}}+h_{\mathrm{pg}}N_{\mathrm{p}}\pi\mathrm{d}^{2}(T_{\mathrm{g}}-T_{\mathrm{p}})A \tag{6.6}$$

where

$C_{\mathrm{p,g}}$ = specific heat capacity for gas phase
$C_{\mathrm{p,p}}$ = specific heat capacity for particle
T_{p} = particle temperature
N_{p} = number of particles in the specified volume = $6\varepsilon_{\mathrm{p}}/\pi\mathrm{d}^3$
d = particle diameter
h_{wg} = convective heat transfer coefficient between the wall and the gas phase
h_{pg} = particle–gas convective heat transfer coefficient

The complementary equations that can be referred to achieve the closure of the balance equations for the gas and particle phases are:

Complementary Equations		
$\tau_{\mathrm{wg}}=\begin{cases}\frac{1}{2}\varepsilon_{\mathrm{g}}\rho_{\mathrm{g}}v_{\mathrm{g}}^{2}\frac{16}{\mathrm{Re_D}} & \mathrm{Re_D}<2100\\ \frac{1}{2}\varepsilon_{\mathrm{g}}\rho_{\mathrm{g}}v_{\mathrm{g}}^{2}\left(0.0791\mathrm{Re_D^{-0.25}}\right) & \mathrm{Re_D}\geq 2100\end{cases}$ where $\mathrm{Re_D}=\rho_{\mathrm{g}}v_{\mathrm{g}}D/\mu_{\mathrm{g}}$	(Blasius formula)	(6.7) (6.8)
$\tau_{\mathrm{wp}}=\frac{1}{2}\varepsilon_{\mathrm{p}}\rho_{\mathrm{p}}v_{\mathrm{p}}^{2}\left(0.0126\varepsilon_{\mathrm{p}}^{-0.021}\varepsilon_{\mathrm{g}}^{-3}\right)$	[6]	(6.9)
$h_{\mathrm{pg}}=\frac{k_{\mathrm{g}}}{d}\left(0.4\,\mathrm{Re_p^{1/2}}+0.06\,\mathrm{Re_p^{2/3}}\right)\mathrm{Pr}^{\,0.4}\left(\frac{\mu}{\mu_{\mathrm{p}}}\right)^{1/4}$ where $\mathrm{Re_p}=\rho_{\mathrm{p}}\lvert v_{\mathrm{g}}-v_{\mathrm{p}}\rvert d/\mu_{\mathrm{g}}$,	[7]	(6.10)
$h_{\mathrm{wg}}=\begin{cases}\frac{k_{\mathrm{g}}}{D}\left(3.66+\frac{0.0688\frac{D}{x}\mathrm{Re_D\,Pr}}{1+0.04\left(\frac{D}{x}\mathrm{Re_D\,Pr}\right)^{2/3}}\right) & \mathrm{Re_D}<4000\\ \frac{k_{\mathrm{g}}}{D}0.023\,\mathrm{Re_D^{0.8}Pr}^{n} & \mathrm{Re_D}\geq 4000\end{cases}$ $n=\begin{cases}0.4 & T_{\mathrm{w}}\geq T_{\mathrm{g}}\\ 0.3 & T_{\mathrm{w}}<T_{\mathrm{g}}\end{cases}$	[8]	(6.11) (6.12)

6.2.2 One-dimensional Breakage Algorithm (ODBA)

The particle breakage modelling using a one-dimensional, two-phase flow model is performed by coupling it with the ODBA. The coupling implementation can be

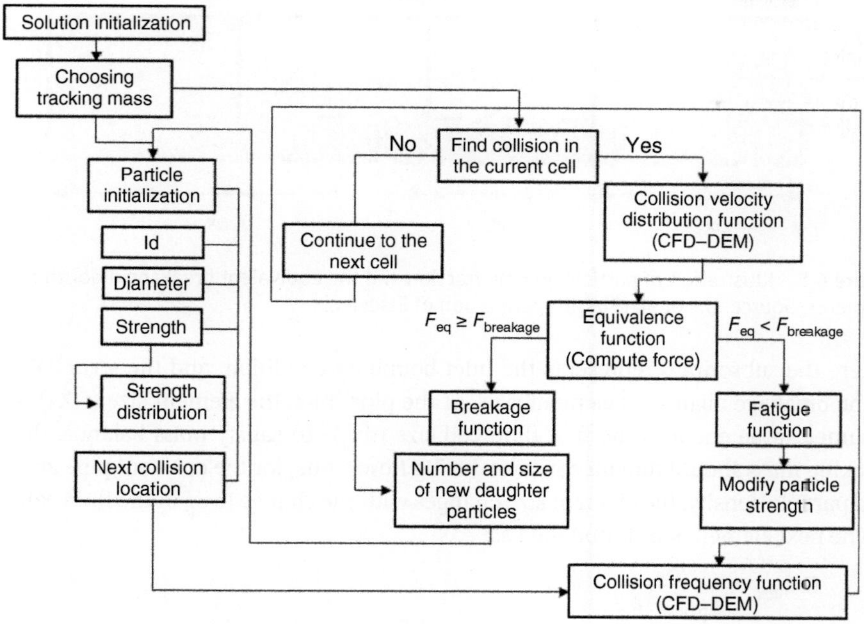

Figure 6.4 Schematic flow diagram of the ODBA.

done either one-way or two-way, based on the choice of not including or including the effect of particle breakage on two-phase flow.

The ODBA starts at inlet with a 'tracking mass' obtained by modifying the known PSD. Further, for the flow of tracking mass along the pipeline, the ODBA accounts the breakage of this tracking mass. To initiate the tracking mass at the inlet, particle information in addition to the size is needed, such as id, next collision time, and strength. The breakage or reduction in the strength of the tracking mass is accounted from the collision search algorithm. In this, the collision time for each particle to collide with another particle and with the pipe wall is obtained from the collision frequency distribution function. For each collision, the selection function is called to choose whether the particle breaks or not. If the particle breaks, the breakage function is called to generate the new fragments having strength based on strength distribution. If the particle is strong enough to survive the impact, the particle will endure fatigue, which will be dictated by the fatigue function. In Figure 6.4, the ODBA concept is summarized in a flow chart.

To understand the algorithm, let us consider a pipe with a two-phase flow (Figure 6.5), which has been solved by the one-dimensional, two-phase flow model. The pipe is divided into four fluid cells, as shown in Figure 6.5. In this, the fluid cells are represented as black lines, whereas the grey area inside the fluid cell represents the particle volume fraction (ε_p).

The tracking mass at the pipe inlet is defined as:

$$m_{n,0} = \varepsilon_{p,0} A \rho_p D X_0 \tag{6.13}$$

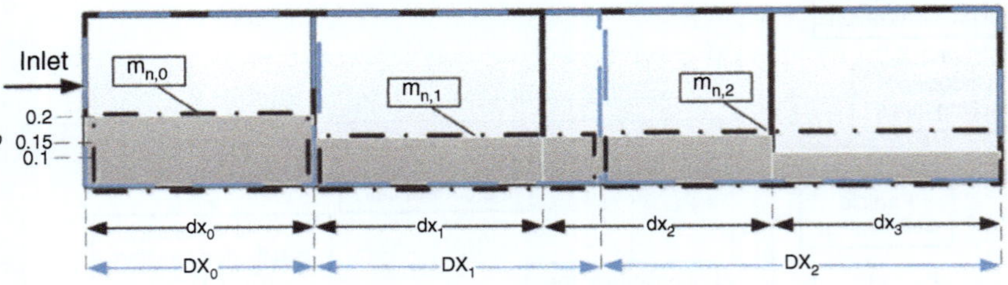

Figure 6.5 Illustration of particle volume fraction and the equivalent breakage algorithm elements. Source: Uzi et al. [9]/With permission of Elsevier.

where the subscript 0 represents the inlet boundary condition, and the term 'DX' is the breakage algorithm element size. At the pipe inlet, the element size (DX_0) is assumed to be equal to the first fluid cell size (dx_0). To satisfy mass balance, the tracking mass should remain constant during flow. Thus, for the constant pipe area and particle density, the element size changes with the change in ε_p in the fluid cell.

The next element size is obtained as:

$$DX_{\text{next}} = \frac{m_n}{\varepsilon_{\text{p,next}} A \rho_{\text{p}}} \tag{6.14}$$

From Eq. (6.14), it can be noted that the reduction in particle volume fraction increases the element size for the constant tracking mass, pipe area, and particle density. Fig. 6.5 demonstrates the change in element size with the change in the particle volume fraction.

The tracking mass, although constant in size, is a sum of the mass of discrete particles.

$$m_{n,0} = \frac{4\pi}{3} \rho_{\text{p}} \sum_{i=1}^{n_{\text{tot}}} d_i^3 \tag{6.15}$$

where n_{tot} is the number of particles in the particular element under consideration.

The time taken by the tracking mass from the entrance to the exit of every element is obtained by translating the resident time of the particles in the continuous domain to the corresponding ODBA elements located in the same positions. The resident time of the particles, for a straight pipe, is known from the one-dimensional flow model by integrating the time from the velocity i.e., $dt/dx = 1/v_{\text{p}}$, in which 'x' is the axial direction of the pipeline system, t denotes time, and v_{p} is the particle velocity. The resident time inside the bend is calculated as: $t = lR/u_{\text{in}}$, where u_{in} is the particle velocity upon entering the bend, R is the bend radius, and l is the equivalent nondimensional bend length that can be approximated as 0.6 for the small radius bend.

In the time interval for the movement of tracking mass in each element, ODBA searches for collisions. After these collisions have been taken into the calculation, the tracking mass is then moved to the next element along the pipeline, situated in DX_1 (as shown in Figure 6.5), with a new particle size distribution and different collision times. Continuing this procedure allows tracking of the same mass along the pipeline at different times and different locations.

The collision search algorithm begins in every element when the tracking mass enters the specific element at the corresponding time. The collisions are searched for every discrete particle until the flow residence time in the element is reached. When a collision is found inside an element, the velocity of the impact is obtained from the statistical impact velocity distribution function (machine function). The equivalent impact force is then calculated from the equivalence function. If the resulting force is stronger than the particle strength, the particle will break. The numbers and sizes of the new daughter particles are obtained from the breakage function. Every new daughter particle is initialized with a new id, diameter, strength, and time of the next collision. In contrast, if the force is weaker than the particle strength, the particle strength will be diminished according to the fatigue function.

By applying the frequency function, the next collision time of every particle is known. Therefore, when a particle travels along the pipeline with a known velocity, it will collide when its traveling time equals the time determined by the frequency function.

6.2.2.1 Machine Functions

Machine functions contribute to particle dynamics in a specific system. They provide information on how often a particle will collide with a wall or with another particle and on the velocity of the impact. To develop machine functions and to find a relationship with the governing parameters, the collision data from numerical modelling can be used.

Uzi and Levy [10] performed CFD–DEM modelling of the pneumatic conveying system to develop machine functions. To analyse the geometrical dependency of machine functions, the conveying system of 1-inch pipe diameter was considered, which is composed of five straight pipes that are connected with four short-radius bends ($r/D = 1$). Figure 6.6 illustrates the considered conveying system. The operating conditions at which the simulations were carried out are shown in Table 6.1.

The collision dynamics of a particle include the frequency of collision, impact velocity, and impact angle. The collision frequency can determine from the ratio of the number of collisions to the total number of particles inside the domain during

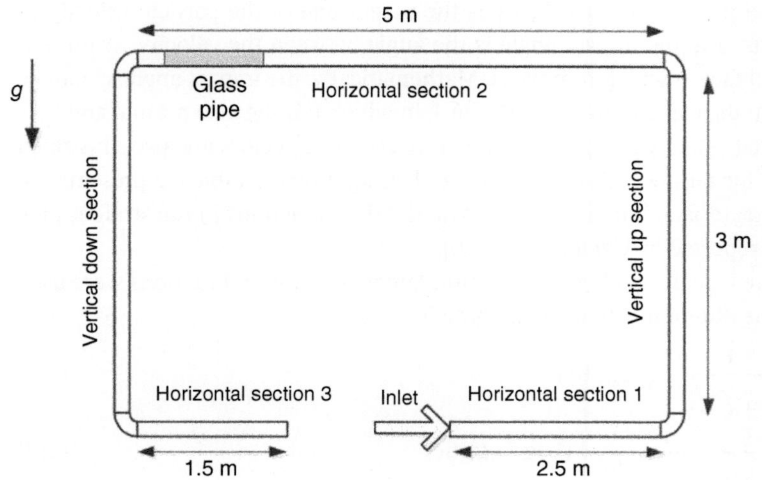

Figure 6.6 Schematic of the pneumatic conveying system adopted for numerical modelling. Source: Uzi et al. [11]/With permission of Elsevier.

Table 6.1 Simulation conditions.

Case number	Velocity ($m\,s^{-1}$)	Material	Particle mass flow rate ($kg\,s^{-1}$)	Gas pressure drop/mass flux
1.1	70	Sand	0.034	$0.0584\,kg\,s^{-1}$
2.1	17	Potash	0.094	80 kPa
2.2	20	Potash	0.094	120 kPa
2.3	29	Potash	0.094	240 kPa

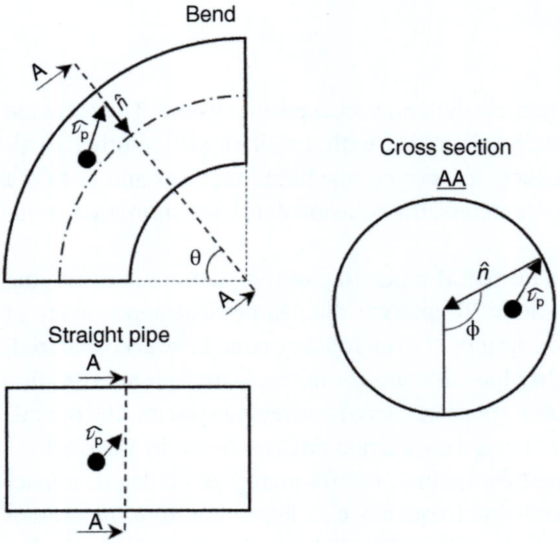

Figure 6.7 Definition of geometric parameters for P–W collision in bend and straight pipe.

a certain time frame. Impact velocity is the magnitude of the particle velocity as collision starts, and the impact angle is the angle between the velocity vector and the numerical face at which it impacts. Mathematically, the impact angle, α, can be obtained from the expression $\alpha = \sin^{-1}(\hat{n}\hat{v}_p)$, in which $\hat{n}$ is the face normal and $\hat{v}_p$ is the normalized velocity vector before the impact. These vectors are also illustrated in Figure 6.7 for a P–W collision in bend and straight pipes. Table 6.2 presents the collision characteristic functions for P–W and P–P interactions in the straight pipe and bend, as reported by Uzi and Levy [10].

In these, the logistic and log-logistic cumulative distribution functions were used to describe the different characteristic functions as:

$$s_i = \frac{1}{1 + e^{\frac{f_i - \mu_i}{\sigma_i}}} \quad \text{(Logistic)} \tag{6.16}$$

$$s_i = 1 - \frac{1}{1 + \left(\frac{f_i}{e^{\mu_i}}\right)^{1/\sigma_i}} \quad \text{(Log-Logistic)} \tag{6.17}$$

Table 6.2 Summary of collision characteristic functions.

Parameter	Distribution type	Parameter	Distribution parameters		Specification
			μ	σ	
Particle-wall interactions					
Circumference angle in bends	Logistic	$s_\phi = \frac{1}{1 - e^{-(\phi-\mu)/\sigma}}$	0	a*	Case 1.1, $d \approx 0.25$ mm
Bend angle	Logistic	$s_\theta = \frac{1}{1 + e^{-(\theta-\mu)/\sigma}}$	53	12	All cases
Impact angle for straight sections	Logistic	$s_\alpha = \frac{1}{1 + e^{-(\alpha-\mu)/\sigma}}$	0	4.315	Case 1.1, $d \approx 0.25$ mm
			0	10.5	Cases 2.1–2.3, $0.5 < d < 1$
			0	8	Cases 2.1–2.3, $1 < d < 2$
			0	5.2	Cases 2.1–2.3, $d\ 2 < d < 5$
Impact angle for bends	Logistic	$s_\alpha = \frac{1}{1 + e^{-(\alpha-\mu)/\sigma}}$	0	14.57	Case 1.1, $d \approx 0.25$ mm
			32.6	10.06	Cases 2.1–2.3, $0.5 < d < 1$
			29.68	9.36	Cases 2.1–2.3, $1 < d < 2$
			25.63	8.79	Cases 2.1–2.3, $2 < d < 5$
Non-dimensional velocity magnitude	Logistic	$s_{\widetilde{v}} = \frac{1}{1 + e^{-(\widetilde{v}-\mu)/\sigma}}$	1	0.125	Case 1.1, $d \approx 0.25$ mm, bends and straight sections
			1	0.18	Bends in cases 2.1–2.3, $0.5 < d < 5$
			1	0.13	Straight sections in cases 2.1–2.3, $0.5 < d < 5$
Non-dimensional frequency	Logistic	$s_f = \frac{1}{1 + e^{-(f-\mu_f)/\sigma_f}}$	b*	d*	Straight sections in cases 2.1–2.3, $0.5 < d < 5$
	Log-Logistic	$s_f = 1 - \frac{1}{1 + \left(\frac{f}{e^{\mu_f}}\right)^{1/\sigma_f}}$	c*	e*	Bends in cases 2.1–2.3, $0.5 < d < 5$

(Continued)

Table 6.2 (Continued)

Parameter	Distribution type	Parameter	Distribution parameters		Specification
			μ	σ	
Particle–particle interactions					
Non-dimensional normal impact velocity	Log-Logistic	$s_{\tilde{v}} = 1 - \frac{1}{1 + \left(\frac{\tilde{v}}{e^{\mu_{\tilde{v}}}}\right)^{1/\sigma_{\tilde{v}}}}$	1	f* and g*	Case 2.1–2.3, $0.5 < d < 5$, bends and straight sections
Non-dimensional frequency	Log-Logistic	$s_f = 1 - \frac{1}{1+\left(\frac{f}{e^{\mu_f}}\right)^{1/\sigma_f}}$	h*	j*	Straight sections in cases 2.1–2.3, $0.5 < d < 5$
			i*	j*	Bends in cases 2.1–2.3, $0.5 < d < 5$

a* $\sigma_\phi = 21.58 - 0.020043\theta^2 + 0.0019994\theta^3 - 6.731 \cdot 10^{-5}\theta^4 + 8.87 \cdot 10^{-7}\theta^5 - 3.958 \cdot 10^{-9}\theta^6$.
b* $\mu_f = 59.83\tau + 5.89$.
c* $\mu_f = 7.06\tau + 3.108$.
d* $\sigma_f = 0.08407\mu_f - 0.50244$.
e* $\sigma_f = 0.154\mu_f - 0.3246$.
f* $\sigma_{\tilde{v}} = 0.1065\tau + 0.34128$ (straight section).
g* $\sigma_{\tilde{v}} = 0.14913\tau + 0.3676$ (bend section).
h* $\mu_f = 15.994\tau + 0.84$.
i* $\mu_f = 13.994\tau + 2.6095$.
J* $\sigma_f = 0.0267e^{0.4833\frac{\mu_f}{SLR}}$.
Source: Adapted from [10].

where the subscript i denotes a specific distribution function, such as collision frequency, collision velocity magnitude, impact angle with the wall surface, circumference impact angle, or bend angle. Further, S_i is a random probability, f_i is the distributed property, μ is the median, and σ is the scale factor of the distribution. Further, the collisional velocity function is described through a nondimensional velocity (impact velocity divided by its median impact velocity, v_{mag}/v_{50}), and the collisional frequency function is described through a non-dimensional frequency expressed as:

$$f* = f\tau \tag{6.18}$$

where τ is the time for a particle to displace itself and expressed as:

$$\tau = \frac{d_p}{v_p} \tag{6.19}$$

in which d_p is the particle diameter in mm and v_p is the particle velocity.

6.2.3 Comminution Functions

In the ODBA, particle characteristics are described by empirical comminution functions. The comminution functions include the following: the initial strength distribution function, the equivalence function, the fatigue function, and the breakage function. The implementation of comminution functions involves converting the probability comminution functions into individual particle properties by a random method and then converting the velocity-dependent comminution functions into strength-dependent ones.

6.2.3.1 Initial Strength Distribution Function

To represent the particle strength distribution, different approaches were adopted over the years, such as the approach based on crushing force – the force at which the particle breaks; based on breakage energy – energy that was applied to break the particle; based on loading stress – the stress required to break the particle; or based on specific energy – breakage energy per particle mass. The representation of particle strength in terms of crushing force is frequently adopted due to its simplicity and higher accuracy. Rozenblat et al. [12] conducted compression tests on different particle types and correlated the particle strength to the crushing force (F_c) using the explicit logistic model, given as:

$$F_c = \left(A_F + B_F \cdot d_{mean}^2\right)\left(\frac{S_F}{1 - S_F}\right)^{1/D_F} \tag{6.20}$$

where S_F is the cumulative probability for crushing, d_{mean} is the particle mean sieve diameter (mm), A_F and B_F are the material constant, and D_F defines the distribution wideness (as D_F becomes larger, the distribution becomes narrower). The empirical constants for different solids are presented in Table 6.3. The initial strength of a particle can be chosen randomly by selecting a random number for S_F between 0 and 1 as an input for the empirical strength (crushing force) distribution.

Table 6.3 Empirical parameters for the crushing force model.

Material	A_F (N)	B_F (N/mm^2)	D_F
Salt	6.7819	3.3203	5
Potash	4.777	2.4298	5
GNP	19.2315	4.4938	6.45
Glass	88.0802	84.9213	7
Sugar	4.6849	2.0667	2.7
Basalt	43.3254	11.3097	2.5
Marble	−5.498	5.3606	3.2

Source: Rozenblat et al. [12]/With permission of Elsevier.

6.2.3.2 Equivalence Function

The equivalence function describes the probability of a particle to break under a certain load or impact velocity. As the particle undergoes impact, the equivalence functions act as translations of the impact velocity to the equivalent force acting on the particle under compression, leading to similar damage. Rozenblat et al. [13] developed an equivalence force model by using the impact test data of particles of different materials and different sizes. It is given by:

$$F_{eq} = A_{eq} \cdot v_{impact}^{B_{eq}} \cdot d_0^{3.3 \cdot B_{eq}} \tag{6.21}$$

where F_{eq} is the equivalent force (N), d_0 is the initial mean particle diameter (mm) of the given size fraction, v_{impact} is the velocity of impact (m/s), and A_{eq} and B_{eq} are the model parameters. The model parameters for different materials are presented in Table 6.4.

For a particle impacting another particle, the equivalent impact velocity is calculated as:

$$v_{pb} = v_{impact} \left(\frac{k_{pp}^4}{k_{pw}^4} \left(\frac{73 + 27(e_{pp} - \gamma_{pp})}{73 + 27(e_{pw} - \gamma_{pw})} \right)^{15} \right)^{1/12} \tag{6.22}$$

where k, and e represent the equivalent stiffness and the coefficient of restitution (COR), respectively, and γ is a factor that depends on COR. The suffixes pp is for

Table 6.4 Equivalence model empirical parameters.

Material	A_{eq}	B_{eq}
Salt	2.94E + 05	0.53
Potash	1.09E + 06	0.65
GNP	1.14E + 06	0.60

Source: Rozenblat et al. [13]/With permission of Elsevier.

particle–particle interaction and pw is for particle–wall interaction. The details to evaluate the equivalent stiffness and factor 'γ' for P-P and P-W interaction can be referred from the literature [14].

6.2.3.3 Fatigue Function

The strength of the particles decreases during repeated impact loading when the load intensity is smaller than the particle strength. This reduction in strength of the particles due to repeated loading is generally named particle fatigue. In pneumatic conveying, the particle collides many times in the system with other particles and with the pipe wall. Thus, the phenomenon of particle fatigue is particularly important for analysing the breakage in the conveying line. Rozenblat et al. [15] developed a fatigue function from the compression test data of the unbroken particles collected after each cycle of impact loading. It is given by:

$$F_i = F_{i-1}\left\{1+\frac{1.89\cdot 10^{13}}{\exp\left[\left(827.8\cdot\left(A_\mathrm{a}\cdot d_0^{B_\mathrm{b}}\cdot F_{i-1}^{C_\mathrm{c}}\right)\cdot F_{i-1}^{-1/3}\right)-\left(A_\mathrm{a}\cdot d_0^{B_\mathrm{b}}\cdot F_{i-1}^{C_\mathrm{c}}\right)\cdot\left(\frac{F_\mathrm{eq}}{F_{i-1}}\right)^{1/3}\right]^2}\right\}^{-3/2} \tag{6.23}$$

where F is the particle strength (N), index 'i' indicates the ith impact of the specific particle and 'i–1' represents the previous impact, F_eq is the equivalence force (N), d_0 is the initial particle diameter (mm), and A_a, B_b, and C_c are the empirical constants whose values reported as 6.724×10^{-3}, –0.725 and 0.333, respectively.

6.2.3.4 Breakage Function

The breakage function describes the mass of the broken particles or particle fragment size distribution in terms of mass, resulting from the breakage of the parent particles. Rozenblat et al. [16] conducted impact experiments with particles of different materials and sizes at different velocities and impact angles to develop a breakage function. The model for breakage function ($B_{i,j}$) for the fraction of the particles that transferred from 'j' size class into, a smaller size, 'i' class is given by:

$$B_{i,j} = 1-(1-t_{10})^{(9/(d_0/D_i)-1)^\alpha} \tag{6.24}$$

where t_{10} indicates the fraction that passes 1/10 of the parent particle size (class j), D_i is the sieve size of fragments in class 'i', d_0 is the parent particle size, and α is the impact angle. The parameters t_{10} and α, are expressed as:

$$t_{10} = 1-\exp\left(-av_\mathrm{impact}^b d_0\right) \tag{6.25}$$

$$\alpha = a_\alpha v_\mathrm{impact}^{b_\alpha} d_0^{C_\alpha} \tag{6.26}$$

Table 6.5 Breakage function empirical parameters.

	$t_{10} = 1 - \exp\left(-av^{b}_{\text{impact}} d_0\right)$		$\alpha = a_\alpha v^{b_\alpha}_{\text{impact}} d_0^{c_\alpha}$		
Materials	**A**	**b**	a_α	b_α	c_α
Salt	5.0×10^{-2}	2.0	0.6	−0.3	−0.24
Potash	2.8×10^{-3}	2.3	1.1	−0.2	−0.15
GNP	1.3×10^{-3}	2.6	0.8	−0.2	−0.22
Zeolite	1.7×10^{-3}	2.6	0.7	−0.2	−0.25

Source: Rozenblat et al. [16]/With permission of Elsevier.

The empirical constants in the model for the parameters t_{10} and α are presented in Table 6.5.

6.2.4 Results and Discussion

The ODBA coupled one-dimensional two-phase flow model has been successfully applied to simulate the particle breakage in the conveying line. A comparison of the model predictions for particle breakage to the experimental data is shown in Figure 6.8. It is obtained while conducting the simulation for 10 conveying cycles at the inlet gas superficial velocity of 40 m/s in the conveying system as shown in Figure 6.6. Every cycle consisted of a single transport of the tracked mass of potash particles of size 2–4 mm through the conveying system, i.e. from the inlet to the filter at the outlet. The filter has sieved the particles under a diameter of 2 mm. After

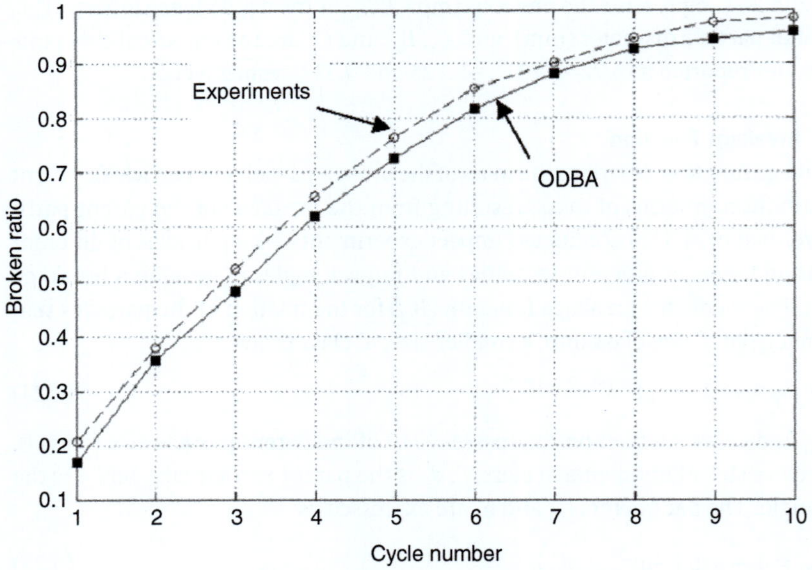

Figure 6.8 Broken ratio of particles after every conveying cycle. Source: Uzi et al. [17]/With permission of Elsevier.

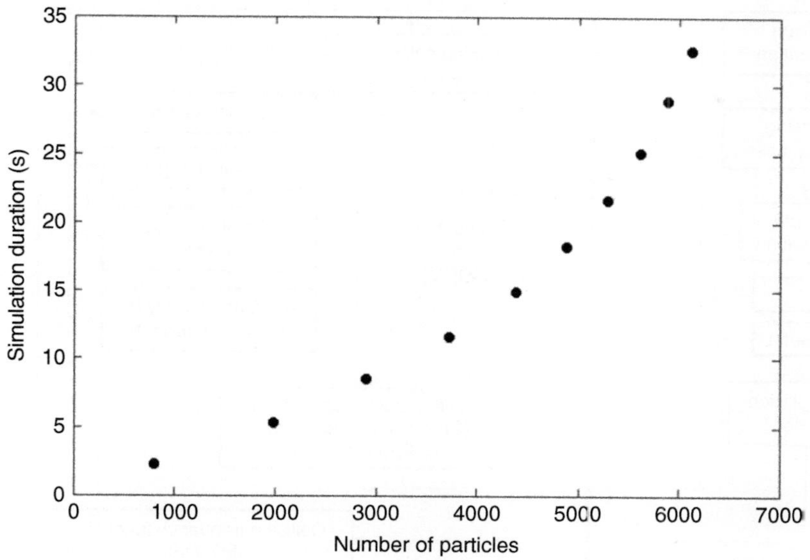

Figure 6.9 ODBA simulation duration in each cycle in terms of particle number. Source: Uzi et al. [9]/With permission of Elsevier.

each cycle the broken ratio, defined as the ratio between the filtered mass and the initial mass, is determined. The predicted variation in broken ratio after every conveying cycle is shown in Figure 6.8. It shows the continuous increase in the broken ratio with conveying cycles. For the 10 conveying cycles, 95% of the particles broke, and this trend would continue to 100% if more cycles were conducted. Further, the predicted results from the ODBA with two-way coupling showed an excellent agreement with the experimental results.

Figure 6.9 presents the simulation time in each cycle in terms of particle number with an ODBA coupled one-dimensional, two-phase flow model. It shows that the calculation time did not exceed 35 sec per cycle, even if the tracking mass contained more than 6000 particles. Thus, the ODBA facilitates a relatively fast calculation time to predict particle breakage and is most suitable to simulate the large-scale conveying systems.

6.3 One-dimensional Surface Erosion Modelling of the Conveying Pipeline

The numerical prediction of surface erosion involves the modelling of particle motion in the process to capture their collision with the target surface, and the calculation of the target material loss from the collision data by using the suitable erosion model.

One-dimensional erosion modelling (ODEM) has followed a similar approach to estimate the useful life of the conveying line based on the erosive wear performance. The ODEM followed the same tracking process as mentioned above for the ODBA.

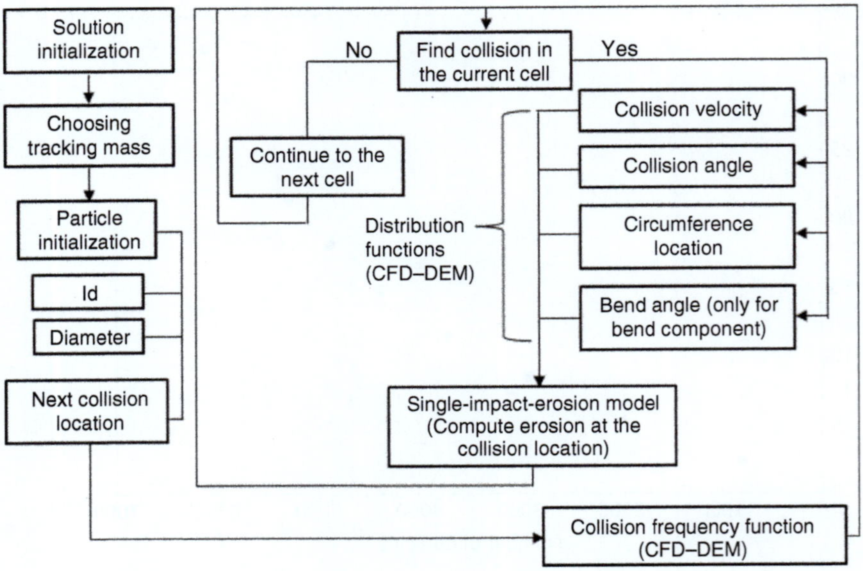

Figure 6.10 Schematic flow diagram of the ODEM.

Moreover, it also utilizes additional machine functions, such as the circumference impact location and the bend angle (see in Table 6.2), because they are needed to dictate the specific impact location on the wall of the conveying system to obtain a full erosion map. The other modification in ODEM from ODBA is that the impact characteristics have been used to evaluate the erosion rate by using the erosion model instead of the breakage scheme. In Figure 6.10, the ODEM concept is summarized in a flow chart.

6.3.1 Erosion Models

Over the years, different erosion equations were reported by the investigators, since the first model of Finnie [18]. Some of these equations were developed based on the experimental tests, considered as empirical models, and the other equations were derived theoretically from the understanding of particle impact characteristics, particle and target material properties, and the erosion mechanisms. The theoretically developed models for metallic and polymeric materials are presented below; moreover, a detailed summary of the available theoretical and empirical models to evaluate the erosion can be found in the literature [19].

For metallic materials [20]

$$\frac{\Delta Q}{m_\mathrm{p}} = C_\mathrm{C}(f+1)[1-\exp(-200\alpha^2)]\left(\frac{\rho_\mathrm{p}}{\eta}\right)^{0.5(1-f)}(d_\mathrm{p})^{(1-f)}V_\mathrm{p}^{3-f}\cos^2(\alpha)\sin^{1-f}(\alpha) + C_\mathrm{D}\left(\frac{\rho_\mathrm{p}^{0.42}}{\eta^{1.25}}\right)(V_\mathrm{p}\sin\alpha)^{2.83} \tag{6.27}$$

For polymeric materials [21]

$$\frac{\Delta Q}{m_{\mathrm{p}}} = C_{\mathrm{C}}(2f+1)[\exp(-\alpha^4)]^{f-0.5}(\eta\rho_{\mathrm{p}})^{0.5(1-f)}(d_{\mathrm{p}})^{2(1-f)}V_{\mathrm{p}}^{3-f}\cos^2(\alpha)\sin^{1-f}(\alpha)$$
$$+ C_{\mathrm{D}}V_{\mathrm{p}}^2\sin^2(\alpha)[1-\exp(-2\alpha^4)]^2 \tag{6.28}$$

where

ΔQ = volume of material removal
m_{p} = particle mass
d_{p} = particle diameter
η = particle shape factor
α = impact angle
C_{C},C_{D},f = model coefficients

The model coefficients in the above equations can be determined by fitting the equations to the experimental erosion data of the particle and target material combination from laboratory testing. The model coefficients for volume loss (mm^3) of Aluminum 6061 target due to sand particle impact, using Wong et al. [22] data, are $f = 0.6$, $C_{\mathrm{D}} = 1.374 \times 10^{-4}$, and $C_{\mathrm{C}} = 3.641 \times 10^{-3}$.

6.3.2 Results and Discussion

The ODEM capabilities to predict the erosion rate of the conveying line has been checked by comparing its predicted results to the CFD–DEM results. Figure 6.11 shows the ODEM and CFD–DEM results correspond to the outer part of the

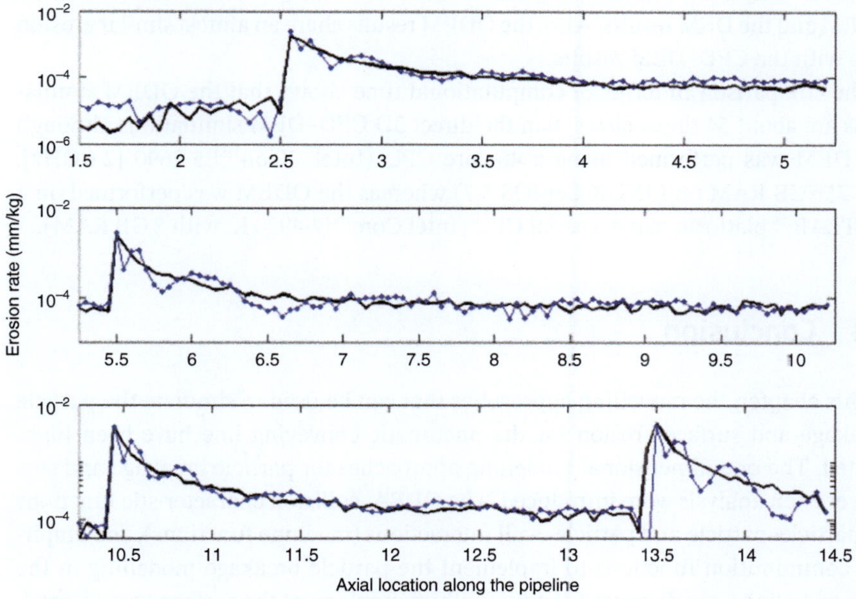

Figure 6.11 Erosion rate of straight sections: blue dotted line for ODEM results and black lines for CFD–DEM results. Source: Uzi et al. [11]/With permission of Elsevier.

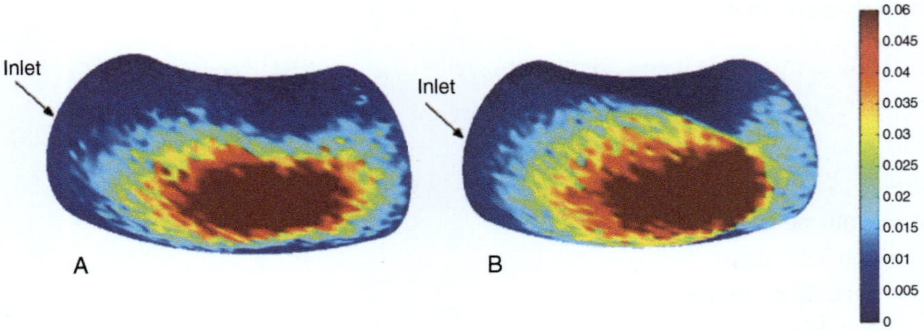

Figure 6.12 Erosion rate contour of the second bend in mm/kg: A – ODEM, B – CFD–DEM. Source: Uzi et al. [11]/With permission of Elsevier.

straight section ($-10° < \phi < 10°$) in terms of the axial distance from the conveying system's inlet (see Figure 6.6). The blue dotted lines are representing the ODEM results, whereas the black lines are for the CFD–DEM results. The operating conditions for simulation are as follows: sand particle diameter = 250 μm, gas mass flux = 0.0584 kg/s, particle inlet velocity = 50 m/s, and mass flux = 0.034 kg/s. The erosion depth is calculated by computing the volume of material loss from the erosion model (Eq. 6.27), calibrated with aluminium 6061 erosion data, and dividing it by the area of the face to which the collision was projected. The comparison shows that the erosion rate assessments of the ODEM are comparable to the calculated DEM erosion rate.

Figure 6.12 shows the erosion rate of the second bend (see Figure 6.6) from ODEM and CFD–DEM. The zone of maximum erosion is at the outer bend wall for both the ODEM and the DEM results. Also, the ODEM results share an almost similar erosion map with the CFD–DEM results.

The comparison in terms of computational time shows that the ODEM simulations are about 54 times faster than the direct 3D CFD–DEM simulations although the DEM was performed using a 48-core CPU (Intel® Xeon™E5-2690 [2.6 GHz], and 256 GB RAM on LINUX CentOS 5.7) whereas the ODEM was performed on a MATLAB™ platform using a serial CPU (Intel Core™i7-4930 K, with 8 GB RAM).

6.4 Conclusion

In this chapter, the modelling approaches that can be used to simulate the particle breakage and surface erosion for the pneumatic conveying line have been highlighted. The one-dimensional modelling approaches for particle breakage and surface erosion analysis were introduced. The ODBA, collision characteristic functions for particle–particle and particle–wall interactions (machine functions), and empirical comminution functions to implement the particle breakage modelling in the conveying line were discussed. Additionally, to implement the surface erosion modelling in the conveying line, the ODEM, machine functions, and erosion models were discussed.

The applications of ODBA and ODEM modelling on an industrial conveying system were shown to illustrate their accuracy in comparison to the experimental or CFD–DEM results. Finally, the computational time required to execute the simulation of a large-scale conveying line using the ODBA and ODEM models has been highlighted. Overall, the one-dimensional particle breakage (ODBA) and erosion modelling (ODEM) approaches showed great potential for the design of an industrial pneumatic conveying pipeline.

References

1 Chapelle, P., Abou-Chakra, H., Christakis, N. et al. (2004). Computational model for prediction of particle degradation during dilute-phase pneumatic conveying: modeling of dilute-phase pneumatic conveying. *Advanced Powder Technology* 15 (1): 31–49.

2 Kalman, H., Rodnianski, V., and Haim, M. (2009). A new method to implement comminution functions into DEM simulation of a size reduction system due to particle-wall collisions. *Granular Matter* 11 (4): 253–266.

3 Brosh, T., Kalman, H., and Levy, A. (2011). DEM simulation of particle attrition in dilute-phase pneumatic conveying. *Granular Matter* 13 (2): 175–181.

4 Han, T., Levy, A., and Kalman, H. (2003). DEM simulation for attrition of salt during dilute-phase pneumatic conveying. *Powder Technology* 129: 92–100.

5 Potyondy, D.O. and Cundall, P.A. (2004). A bonded-particle model for rock. *International Journal of Rock Mechanics and Mining Sciences* 41: 1329–1364.

6 Lech, M. (2001). Mass flow rate measurement in vertical pneumatic conveying of solid. *Powder Technology* 114 (1): 55–58.

7 Whitaker, S. (1972). Forced convection heat transfer correlations for flow in pipes, past flat plates, single cylinders, single spheres, and for flow in packed beds and tube bundles. *AIChE Journal* 18: 361–371.

8 Levy, A., Mason, D.J., Levy-Hevroni, D. et al. (1998). Drying of wet solid particles in a steady-state one-dimensional flow. *Powder Technology* 95 (1): 15–23.

9 Uzi, A., Kalman, H., and Levy, A. (2016). A novel particle attrition model for conveying systems. *Powder Technology* 298: 30–41.

10 Uzi, A. and Levy, A. (2018). Collision characteristics for predicting particle attrition and pipe wear in conveying systems. *Chemical Engineering & Technology* 41 (8): 1511–1523.

11 Uzi, A., Ben-Ami, Y., and Levy, A. (2017). Erosion prediction of industrial conveying pipelines. *Powder Technology* 309: 49–60.

12 Rozenblat, Y., Portnikov, D., Levy, A. et al. (2011). Strength distribution of particles under compression. *Powder Technology* 208 (1): 215–224.

13 Rozenblat, Y., Levy, A., Kalman, H. et al. (2013). Impact velocity and compression force relationship-equivalence function. *Powder Technology* 235: 756–763.

14 Brosh, T., Kalman, H., Levy, A. et al. (2014). DEM–CFD simulation of particle comminution in jet-mill. *Powder Technology* 257: 104–112.

15 Rozenblat, Y., Levy, A., Kalman, H. et al. (2013). A model for particle fatigue due to impact loads. *Powder Technology* 239: 199–207.

16 Rozenblat, Y., Grant, E., Levy, A. et al. (2012). Selection and breakage functions of particles under impact loads. *Chemical Engineering Science* 71: 56–66.

17 Uzi, A. and Levy, A. (2017). Particles' degradation and dynamics in conveying systems. *Powder Technology* 311: 247–256.

18 Finnie, I. (1960). Erosion of surfaces by solid particles. *Wear* 3 (2): 87–103.

19 Tarodiya, R. and Levy, A. (2021). Surface erosion due to particle-surface interactions – A review. *Powder Technology* 387: 527–559.

20 Ben-Ami, Y., Uzi, A., and Levy, A. (2016). Modelling the particles impingement angle to produce maximum erosion. *Powder Technology* 301: 1032–1043.

21 Tarodiya, R. and Levy, A. (2021). Erosion of polymers and polymer composites surfaces by particles. *Advanced Powder Technology* 32: 3149–3159.

22 Wong, C.Y., Solnordal, C., Swallow, A. et al. (2012). Predicting the material loss around a hole due to sand erosion. *Wear* 276: 1–15.

7

Discrete Element Modelling of Pharmaceutical Powder Handling Processes

Chao Zheng, Mohammadreza Alizadeh Behjani, Jiawei Hu, and Chuan-Yu Wu

University of Surrey, School of Chemistry and Chemical Engineering, Stag Hill, Guildford, GU2 7XH, UK

Abbreviations

API	active pharmaceutical ingredient
BPM	bonded particle model
CBP	coefficient of blending performance
CFD	computational fluid dynamics
COPD	chronic obstructive pulmonary disease
CoV	coefficient of variation
DoE	design of experiment
DEM	discrete element method
DPI	dry powder inhaler
EMA	European Medicines Agency
FDA	Food and Drug Administration
GPU	graphic processing unit
JKR	Johnson Kendal Roberts
MVRA	multivariate regression analysis
NIR	near-infrared spectroscopy
PEPT	positron emission particle tracking
RSM	response surface methodology
RTD	residence time distribution
SDG	sustainable development goals
TSG	twin screw granulation
VDW	van der Waals

7.1 Introduction

More than two-thirds of medicines in the global pharmaceutical market are in the form of tablets, capsules, and dry powder inhalers (DPIs), which are the so-called solid dosage form and involve the processing and handling of dry powders.

Simulations in Bulk Solids Handling: Applications of DEM and other Methods, First Edition.
Edited by Don McGlinchey.

Most pharmaceutical products consist of one or more active pharmaceutical ingredients (APIs) and various inactive ingredients, also known as excipients, of different functionalities. Typical manufacturing processes consist of a series of powder handling unit operations, such as milling, mixing, blending, powder conveying, granulation, die filling, and compaction. In these unit operations, the bulk behaviours of powders are governed by the dynamics of individual constituent particles. Hence, equipment design, particle properties (e.g. size, shape, and mechanical properties), and process conditions can significantly affect bulk behaviour, manufacturing performance, and properties of the final product.

Given the massive quantities produced globally and high value-added nature of these products, any improvement in manufacturing efficiency could create significant economic (e.g. higher production efficiency and lower costs), societal (e.g. cheaper products with high quality and a healthier nation), and environmental benefits (e.g. lower energy consumption and pollutant emission). It could make a positive contribution to meet the UN sustainable development goals (SDG) [1], such as Goal No. 9 – industry, innovation, and infrastructure, and Goal No. 12 – responsible production and consumption. Therefore, considerable efforts have been made in understanding, predicting, and innovating the powder handling processes and equipment for pharmaceutical products. In particular, mechanistic modelling of various unit operations in pharmaceutical product manufacturing using the discrete element method (DEM) has been increasingly performed to enhance process understanding and to optimise the formulations and manufacturing processes over the last four decades.

The DEM is a numerical method that treats powders as assemblies of individual particles and solves the Newton's second law of motion for each constituent particle, from which not only the dynamics of individual particles but also the bulk powder behaviour can be explicitly determined. It is an effective numerical approach to link the particle properties to the process performance of powders. DEM was first developed by Cundall and Strack in 1979 for geotechnical applications [2] and has ever since been advanced by many researchers and applied to analyse a wide range of problems involving granular materials, such as packing, chute flow, hopper flow, mixing/blending, segregation, agglomeration, fracture/breakage, and compaction.

The application of DEM in the pharmaceutical field dates back to the 2000s and the early studies focused upon the feasibility of the application of DEM in modelling various unit operations in the pharmaceutical industry. For example, Nakamura et al. [3] analysed the motion of tablets in the coating operation using DEM and examined the impact of tablet size and process conditions (i.e. tablet loading and rotation speed) on coating performance. Fu et al. [4] examined the 3D packing structures of model pharmaceutical powders using DEM and compared the numerical results with the structural measurements obtained by X-ray computed tomography. They found that DEM was able to predict the 3D structure of real parking systems. Furukawa et al. [5] analysed the segregation behaviour of binary mixtures with different sized particles during linear die filling using DEM and found that the percolation of fine particles between course particles is the dominant segregation mechanism. Hildebrandt [6] simulated the die filling process in an actual lab-scale rotary tablet press (Fette 1200i, Fette Compacting GmbH, Schwarzenbek, Germany) using DEM and investigated the effect of process conditions and material properties on the fill mass. Yeom and Choi [7] modelled the mixing process in laboratory-scale V-blenders using DEM and explored the feasibility to develop a process scale-up strategy based upon DEM simulations.

With the advances in DEM methodologies and computing hardware, especially high-performance computing with graphic processing units (GPUs) and other parallel computing techniques, DEM has been increasingly employed in modelling pharmaceutical manufacturing processes in the last decade to enhance process understanding, and optimise the formulations and processes. Furthermore, the use of DEM in modelling continuous manufacturing gains considerable momentum recently, as continuous manufacturing enables more flexible and cost-effective production, better process control, and subsequently better quality of drug products, which is well recognised by pharmaceutical companies and encouraged by the regulatory authorities, including European Medicines Agency (EMA) in the EU and Food and Drug Administration (FDA) in USA, but there is a lack of knowledge and predictive tools for continuous manufacturing, which significantly inhibits the wide implementation of continuous manufacturing in the pharmaceutical industry. DEM is regarded as a useful tool to quantitatively analyse these processes.

In this chapter, the application of DEM in modelling the powder process in pharmaceutical product manufacturing is discussed. The fundamentals of DEM are first introduced. An overview of its application in various unit operations is then presented, which includes powder flow, mixing/blending, and milling. Furthermore, recent developments in modelling continuous manufacturing using DEM, such as continuous powder blending and twin-screw granulation, and dry powder inhalation by DEM are also introduced.

7.2 Discrete Element Method

DEM considers a powder as a system made of interacting individual particles and explicitly analyses the motion of each constituent particles. Various particle interactions and complex particle shapes can be considered in DEM; hence, the movements of individual particles in a complex powder handling process can be determined.

7.2.1 DEM principles

The basic principle of DEM contains two main steps, i.e. determining the motion of each particle and modelling the contact force based upon contact models, as depicted in Figure 7.1. In each cyclic calculation, the particle positions are updated according to the velocity and acceleration, which is calculated using second Newton's second law of motion, and new particle positions are then used to determine the contact forces acting on the contact particles.

Generally, two approaches are available to model the particle interactions, namely the hard-sphere and soft-sphere approaches. In the hard-sphere approach, the particles exchange momentum through collisions, i.e. no deformation in particles is

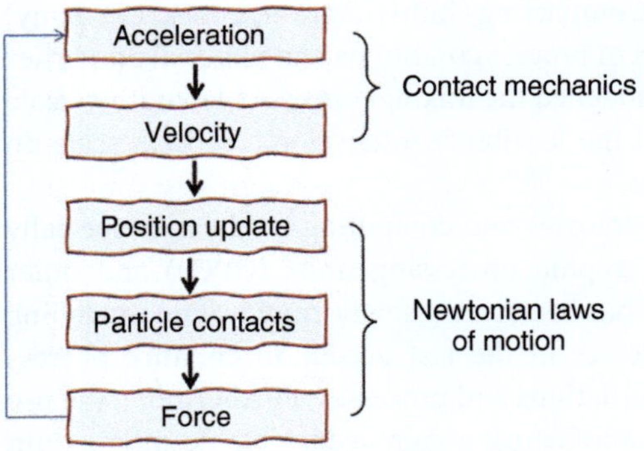

Figure 7.1 Flowchart of particles' dynamics calculations in DEM modelling.

considered. With the soft-sphere approach, the particles are allowed to deform and the interparticle forces are determined according to certain contact models.

In the DEM, particles are generally treated as rigid bodies that can have translational and rotational motions. Therefore, every particle is analysed with six degrees of freedom (three for translations and three for rotations) in a 3D DEM simulation. The second Newton's law of motion is used to calculate the translational and angular accelerations of individual particles. The relationships for the translational and rotational accelerations are expressed in Eqs. (7.1) and (7.2), respectively.

$$m\frac{\mathrm{d}V}{\mathrm{d}t} = mg + F_{\mathrm{c}} + F_{\mathrm{nc}} \tag{7.1}$$

$$I\frac{\mathrm{d}\omega}{\mathrm{d}t} = M \tag{7.2}$$

where m denotes the mass of the particle, V shows the translational velocity, t is the time, mg is the gravitational force applied on the particle, and F_{c} and F_{nc} denote the contact and non-contact forces between the particle and its neighbouring particles or walls. ω is the rotational velocity, I denotes the moment of inertia, and M is the resultant torque exerted on each particle. The interactions of particles and the pertinent forces are schematically presented in Figure 7.2.

To solve the equations of motion and update the velocities and positions accordingly, the accelerations are integrated over a time step, dt. Different methods could

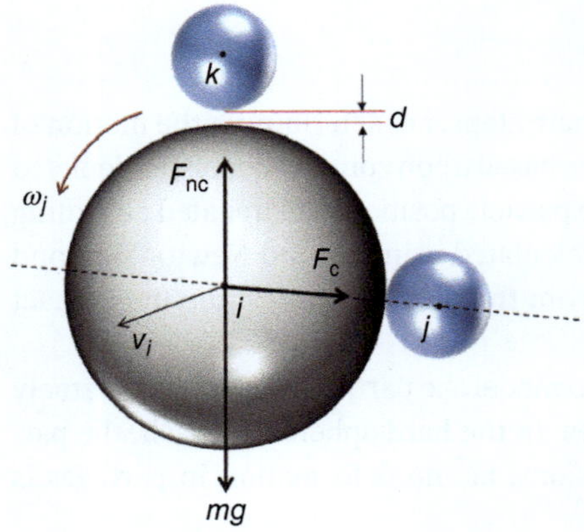

Figure 7.2 Schematic of particles interacting with each other in DEM.

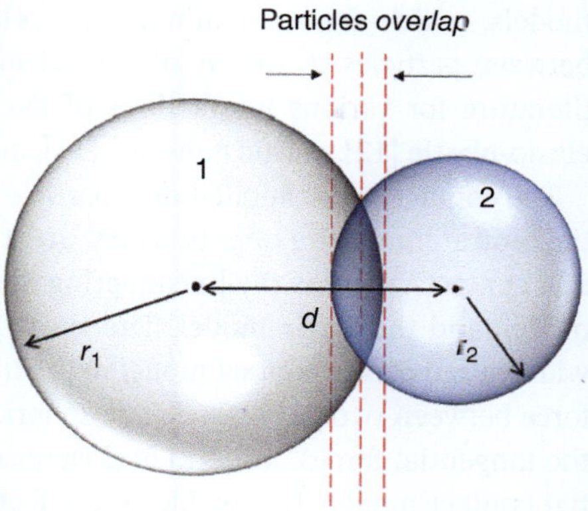

Figure 7.3 Schematic of the artificial overlap between the contacting particles in DEM.

be used for integration in DEM, such as the Verlet algorithm, forward Euler, and frog-leap methods. In DEM, it is assumed that particles come into contact only at the contact points where some overlaps can occur [2], as illustrated in Figure 7.3. The overall movement of particles, including their acceleration, velocity, and positions, are mathematically updated at each time step. Hence, the time step determines the resolution of calculations in DEM and must be selected sufficiently small to avoid the excessive overlap of particles during their interactions with each other. In addition to the forces and torques, disturbance waves coming from far neighbours can impact the motion of a particle. The time step must be small enough so that particle displacement calculations are performed before the disturbance can propagate from a particle further than its immediate neighbours [8]. The criteria for the selection of the time step in the DEM modelling is the Rayleigh surface wave propagation speed. Rayleigh waves are surface waves that travel close to the surface of solids and their speed depends on the material properties and structure of the solid medium. In DEM, the time step should be smaller than the Rayleigh time step (T_R), for example, 20–40% of the Rayleigh time step depending on the application. The Rayleigh time step is expressed in Eq. (7.3).

$$T_R = \frac{\pi R \left(\frac{\rho}{\tau}\right)^{0.5}}{0.1631\nu + 0.8766} \tag{7.3}$$

where R, ρ, τ, and ν show the radius, density, shear modulus, and Poisson's ratio of the particle, respectively. It should be noted that when an assembly of particles with different material properties or size distribution is modelled, the smallest time step should be considered in calculations [9], which can significantly impact the computational time. For example, fine particles require much smaller time steps in the calculations as compared to the coarse ones with the same material properties.

In DEM modelling, forces are categorised into two main classes of contact and non-contact forces. Contact forces are the result of the direct physical contact/collision of particles with their neighbours and the geometry boundaries, while the non-contact forces do not require physical interactions for affecting the particles, such as the electrostatic, magnetic, and gravitational forces. Contact forces can be modelled based upon the principles of contact mechanics, i.e. contact models, which are the mathematical relationships describing the contact forces between particles. A variety of contact models are developed and reported in the literature for various interactions of particles, such as perfectly elastic [10, 11], elasto-plastic [12], elastic adhesive [13], and elasto-plastic adhesive contacts [14].

The contact forces applied on a particle can be summarised into normal, tangential, and rolling components. There are a variety of mathematical models for normal contacts, such as the linear spring dash-pot model, the hysteretic linear spring model, and the Hertz model. The Hertz model is probably the most famous and widely used elastic contact model in the literature [10] which calculates the normal force between two perfectly elastic spherical particles in contact. For calculation of the tangential forces, Mindlin and Deresiewicz [15] proposed the classical tangential contact model. To consider the effect of adhesion between particles, the JKR model [13] can be used. A detailed description of the DEM principles and various contact models can be found in [16, 17].

7.2.2 DEM Calibration

Similar to many other numerical methods, DEM modelling also involves the determination of many input parameters. Some of the input parameters can be readily measured experimentally, such as particle size distribution, density, and particle shape. Some parameters such as particle stiffness, coefficient of static friction, and adhesion are still experimentally obtainable for some materials but require more effort and sophisticated instruments. Other parameters such as the coefficient of rolling friction and Poisson's ratio cannot be obtained directly from experimental measurements, especially for most particles handled in the pharmaceutical industry and hence need to be calibrated. It is now well recognised that the main challenge in the industrial application of DEM is how to calibrate the input parameters [18] so that the simulation is performed specific to certain processes and materials. To select appropriate DEM input parameters, model parameter calibration is necessary. Various calibration approaches are hence proposed and reported in the literature [19, 20]. Nevertheless, calibrating all DEM parameters simultaneously is not methodologically feasible and could be extremely challenging. Thus, in previous studies reported in the literature, most input parameters are kept constant and obtained experimentally or from literature; therefore, only a limited number of parameters are calibrated.

In general, two approaches are proposed to calibrate the DEM input parameters. The first approach focuses on the experimental measurement of a bulk property or response of the powder, e.g. the angle of repose, in a typical powder handling process, and the same process is then modelled using DEM. The DEM input parameters are varied iteratively until the bulk property determined from DEM simulations matches the experimental measurement. This iterative approach relies on the trial-and-error process to approximate the input parameter and is limited to calibration of one parameter at a time. If the number of input parameters to be calibrated up to three and more, this method can become cumbersome due to the large number of trials that are required for the calibration study. Nevertheless, for the calibration of one or two parameters, this approach is reasonable and relatively straightforward.

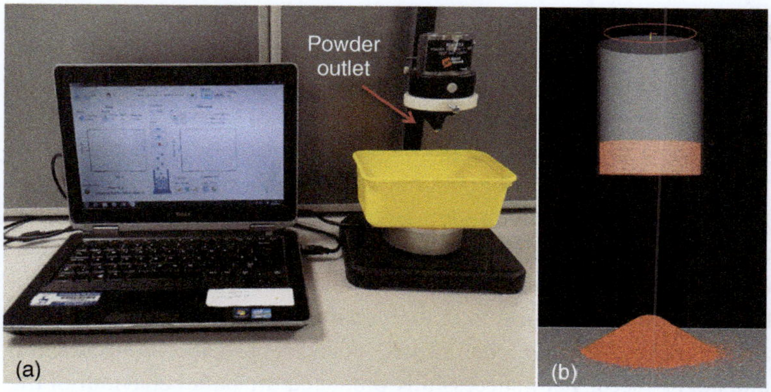

Figure 7.4 The powder discharge process from (a) Flodex device and (b) DEM modelling.

As an example, to calibrate the rolling friction coefficient of a model pharmaceutical powder using this approach, such bulk response as the discharge behaviour can be measured experimentally using a simple device as shown in Figure 7.4a, in which the powder is discharged from the orifice, and the discharged mass is measured versus time. The same procedure can be then modelled using DEM (see Figure 7.4b), and the results are compared with the experiments.

The powder mass discharged is plotted against the time in Figure 7.5. As illustrated in the figure, the rolling friction coefficient is tuned so that the DEM results match the experimental data as closely as possible. The average deviation of the DEM results from the experiment is minimum (i.e. 1.2%) at rolling friction of $C_r = 0.12$; therefore, this value is selected as the calibrated rolling friction coefficient for the DEM simulations.

The second calibration approach employs more advanced techniques such as machine learning [21] and design of experiments (DoE) [22] to systematically select the parameters that best describe the bulk behaviour of the powder. For instance, Boikov et al. [21] suggested using the Random forest algorithm to determine the

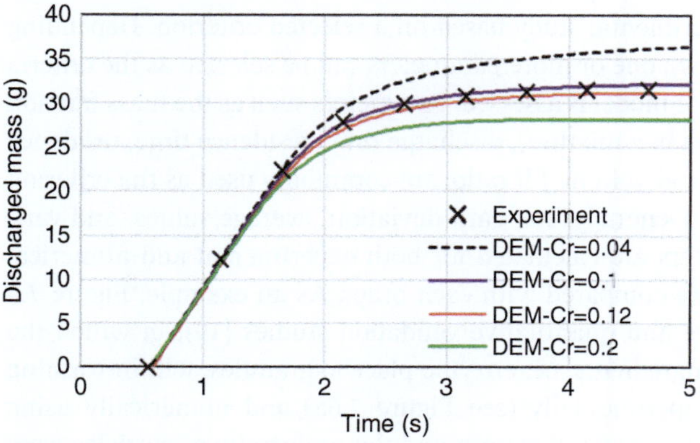

Figure 7.5 Comparison of the powder discharge time obtained from the experimental measurements and DEM modelling.

sensitivity of the measured bulk response to the DEM input parameters. The Random forest is an algorithm used in the machine learning to facilitate the process of decision-making. Boikov et al. [20] used a technical vision method to analyse the images obtained from 200 simulations and used this data set to assess their predictive algorithm. They argued that it was possible to develop a universal method for the calibration of the DEM parameters using this algorithm. El Kassem et al. [22] proposed a systematic calibration method based on combining the DoE and multivariate regression analysis (MVRA) techniques. They used a vertical auger dosing calibration device to obtain three bulk responses and performed a set of DEM simulations by varying four input parameters, i.e. particle–particle and particle-wall sliding friction coefficients, particle–particle rolling friction coefficient, and auger speed. They successfully predicted the bulk behaviour of the powder within the ranges of input parameters using the MVRA technique. As a result, they identified an optimised set of calibrated input parameters whereby their DEM simulations could closely match the experiments. When it comes to complex systems with several input parameters, following an automated approach, e.g. the machine learning, for the calibration of the DEM input parameters is definitely advantageous over the iterative approach. However, a standardised and universal methodology for calibration of the DEM input parameters is still lacking and worthy pursuing.

7.2.3 DEM Validation

Although DEM has been advanced dramatically in the past few decades and its robustness is well recognised, it is important to validate the DEM simulations to ensure that the numerical results are sufficiently accurate and meaningful. Generally, two types of DEM validation can be performed: quantitative validation and qualitative validation. Qualitative validation relies on visualisation and comparison of the particle profiles and flow patterns of powders that can be obtained from DEM modelling and experiments. On the other hand, quantitative validation is referred to the analysis and comparison of the numeric data obtained from the DEM modelling and experimental study based on a selected criterion. Depending on the nature of the process, one or more parameters can be selected as the criteria and the predictability of the model is assessed. Parameters such as the mass fraction (e.g. particle concentration in a mixture), discharge time, residence time, residence time distribution (RTD), and volume fill ratio, are commonly used as the criterion in the validation studies. Generally, standard deviation, average values, and variabilities of these parameters are calculated for both experimental and numerical studies, and the results are compared with each other. As an example, Figure 7.6 illustrates both qualitative and quantitative validation studies [19], in which the segregation of low-level ingredients, i.e. enzyme placebo granules, within washing powders was analysed experimentally (see Figure 7.6a) and numerically using DEM (Figure 7.6b). Both the packing pattern and the uniformity of particles were qualitatively and quantitatively compared to validate the DEM model, respectively. The coefficient of variation (CoV) of the enzyme placebo granules' mass fractions

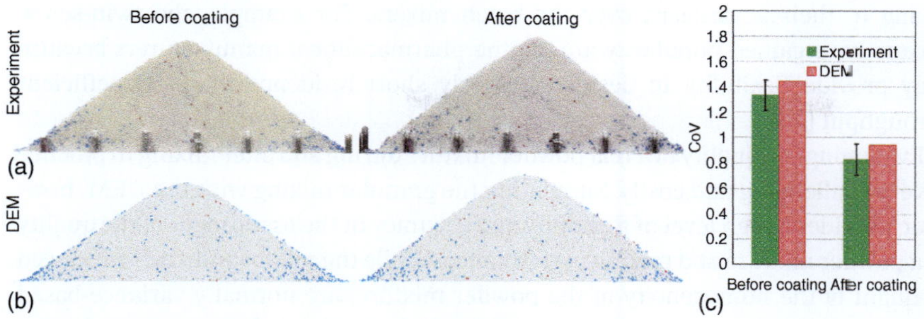

Figure 7.6 Qualitative and quantitative validation of the DEM simulation of granular segregation during the heap formation of a model washing powder. The distribution of the enzyme placebo granules (red particles) and bleach particles (blue particles) on the front surface of the heap is presented for (a) experiments and (b) DEM simulations, before and after coating the enzyme granules. Source: Alizadeh et al. [19]/ with permission of Elsevier.

in the heap was calculated for the experiments and DEM simulations, and a good match between the experimental and numerical results was observed, as shown in Figure 7.6c.

7.3 DEM Modelling of Powder Mixing/Segregation

Pharmaceutical formulations are normally made of various types of powders with different chemical and physical properties. These ingredients must be uniformly mixed in the manufacturing process. Therefore, powder mixing is always regarded as a critical unit operation in the pharmaceutical industry. However, the quality of powder mixtures can always deteriorate over time in the process through a phenomenon called segregation, which has been considered as a serious issue in the pharmaceutical industry. The occurrence of segregation in pharmaceutical manufacturing can lead to under-dose or over-dose of the active ingredients and a large variation in product quality as well as subsequent drug delivery performance. This will lead to failure of the product in quality control, and waste of resources [23].

7.3.1 DEM Modelling of Powder Mixing

Powder mixing is a pivotal process in producing high-quality pharmaceutical products. Understanding the underlying mechanisms of mixing facilitates the design and optimisation of high-performance and reliable mixers. It is shown that the DEM is one of the robust numerical methods for gaining deeper insight into powder mixing. Hence, DEM is widely used to model the mixing process a variety of batch and continuous blenders/mixers, e.g. the tumbling blender [24], twin-screw mixer [25], simple-rotating drum [26], paddle mixer [27], bin blender [28], v-blender [29], agitated tubular mixer [30], tote blender [31], ribbon mixer [32], and slant cone mixer [33]. Among the different types of powder mixing equipment used in the pharmaceutical industry, continuous mixers/blenders receive more attention owing to their advantages over the batch mixers. For example, the twin-screw mixers have gained popularity among the pharmaceutical manufacturers because they provide flexibility in design, relatively short residence time, and efficient throughput [25].

Evaluating the quality of a real powder mixture during and after mixing in practice is very challenging and costly. Simulating the granular mixing with the DEM, however, provides a high level of flexibility and accuracy in the assessment of the quality of a powder mixture and mixing performance. While the mixing indices, introduced to quantify the homogeneity of the powder mixture, are normally variance-based and depend on the sample size, i.e. the value of the mixing index may vary with the resolution and number of the samples taken from the powder mixture. In contrast, DEM provides the exact number and position of the particles at different simulation time periods, which is impossible to obtain experimentally. For example, Behjani et al. [25] proposed a sample-independent mixing index, termed the coefficient of blending performance (CBP), to monitor the formation of the unwanted API agglomerates in continuous blending processes. To obtain CBP, the number of contacts of API-excipient (N_{Ae}) and API-API (N_{AA}) are obtained, and their contact ratios, i.e. the number of contacts over the total number of API or excipient particles, are then calculated using Eqs. (7.4) and (7.5).

$$CR_{\mathrm{Ae}} = \frac{N_{\mathrm{Ae}}}{N_{\mathrm{e}}} \tag{7.4}$$

$$CR_{\mathrm{AA}} = \frac{N_{\mathrm{AA}}}{N_{\mathrm{A}}} \tag{7.5}$$

where CR_{Ae} and CR_{AA} are the contact ratios of API-excipient and API–API, respectively. N_{e} and N_{A} are the total number of API and excipient particles, respectively. CBP is defined as the API-excipient contact ratio over the API–API contact ratio as follows

$$\mathrm{CBP} = \frac{CR_{\mathrm{Ae}}}{CR_{\mathrm{AA}}} = \frac{N_{\mathrm{Ae}}}{N_{\mathrm{AA}}}\frac{N_{\mathrm{e}}}{N_{\mathrm{A}}} \tag{7.6}$$

This index can be determined for mixtures of monosized particles or of certain size distributions. The evaluation using such an index is only possible with the information on all particles at every time step, which can be readily achieved with DEM simulations.

7.3.2 DEM Modelling of Segregation

Segregation, also known as de-mixing, is a phenomenon through which particles separate from each other due to their different physical and mechanical attributes. The main contributing factors for particles to segregate can be divided into two categories: (i) differences in particle attributes and (ii) process dynamics. Difference and variation in particle size, density, shape, surface conditions, and electrostatic properties are among the most important factors leading to segregation [34]. Segregation can be caused through percolation, sifting, projection, and elutriation mechanisms

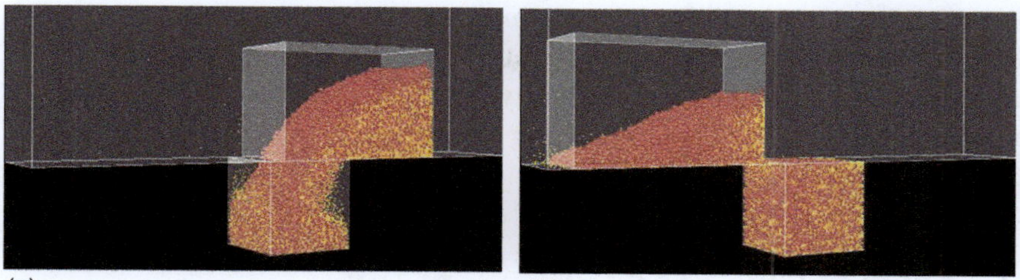

Figure 7.7 Die filling of a binary mixture of large particles (shown in red) and small particles (shown in yellow): (a) during fie filling and (b) after die filling. Source: Guo et al. [40]/with permission of Elsevier.

induced by process dynamics such as vibration [35], shear strain [36], and fluid drag [37], as well as equipment design [38].

DEM is a useful tool for modelling and predicting the segregation tendency of particles during manufacturing processes. Among the particle properties that induce segregation, variability in particle size is probably the most significant and the most widely observed phenomenon. Thus, size-induced segregation is extensively investigated using DEM [39, 40]. For example, Guo et al. [40] numerically explored size induced segregation during die filling and found that significant segregation could be induced due to the difference in particle sizes (see Figure 7.7).

The difference in particle shape also leads to variation in the powder flow behaviour, which can eventually affect the uniformity of the powder mixture. Shape-driven segregation is also widely observed in the industry; however, assessing its significance along with the other factors, such as size and density variation is difficult. DEM analysis is also performed to understand the underlying mechanisms of the shape-driven segregation [41, 42]. DEM simulations have shown that particle shape can cause segregation during the heap formation of binary and ternary powder mixtures [19, 42]. For instance, Alizadeh et al. [42] used DEM to study the effect of particle shape on the segregation tendency of washing powders. They used clumped spheres to mimic the real shape of the particles and showed that some model simplifications, e.g. using spherical particles with augmented rolling friction coefficient lead to underestimation of the extent of the segregation. On the other hand, tailoring particle shape can improve the homogeneity of powder mixtures, as shown by Asachi et al. [43].

Segregation can also be induced due to the difference in particle density. It is widely observed that when building a heap with a mixture of particles of different densities, the particles of low density tend to cascade down the heap free surface and congregate near the base of the heap, while the particles of high density penetrate the heap near the peak (i.e. the pouring point) [44]. As most powder handling processes take place in the presence of air, such as die filling and silo charging, significant segregation can take place due to the difference in particle density. Zeilstra et al. [45] argued that two mechanisms dominates density-induced segregation: (i) acceleration difference in from the air drag and (ii) inertia difference during particle collisions. To model the air, Guo et al. [46] employed a coupled

DEM–CFD, in which the air is modelled using CFD and particles are modelled using DEM, to analyse density-induced segregation during the die filling. They found that segregation could be indued by density, but the degree of segregation depends upon the flow patterns during die filling as well as the significance in density difference.

7.4 DEM Modelling of Continuous Blending

Powder blending is essential in various industries, including food, construction materials, and pharmaceuticals [47]. Especially in the pharmaceutical industry, the powder blending process can significantly impact final product quality, such as mass variation and content uniformity. Various types of blenders are designed for pharmaceutical applications. Moreover, the pharmaceutical industry sees a rapid transition from traditional batch blending to continuous blending, which is beneficial for large-scale production [48]. However, it is very challenging to understand the mixing mechanism partly due to the diverse material properties and vast operation parameter space, which significantly affect the blending process. A series of advanced experimental techniques, such as positron emission particle tracking (PEPT) and near-infrared spectroscopy (NIR), were developed to study the blending process [49]. These experimental techniques are employed to provide more details for better understanding the mixing mechanism and collect a larger amount of data for building a data-based model to predict the production. It is well known that all these techniques require expensive and specialised equipment and stringent operation procedures, and can only access limited information on the continuous blending processes. Consequently, numerical techniques, such as DEM, continuum modelling, and data-based methods, are developed to explore the mixing mechanism and blending performance.

Since powders in the continuous blending processes experience complex flow patterns that are dominated by the collective dynamic behaviour of individual particles, DEM is the most suitable method for analysing continuous manufacturing among the abovementioned numerical approaches. Due to the huge number of particles (a minimum of tens of millions) generally involved in the continuous blending, the early DEM analysis of continuous blending focused on the use of the reduced and compartmental models [50, 51], in which the blender is divided into periodic slices, and the powder flow behaviour in each slice was modelled separately. Sarkar et al. [51] explored the continuous mixing with periodic DEM approach and found that the adopted approach can mimic the powder flow in the central region of the mixer well, but obtained large variances at the inlet or outlet. The reduced or simplified approaches significantly save the computing time; however, it still has limitations to obtain the bulk behaviours at the system level. Previous studies also pointed out that the construction of compartmental models does need a very good understanding of the system. Moreover, great effort is needed to build and optimise the compartmental models [52].

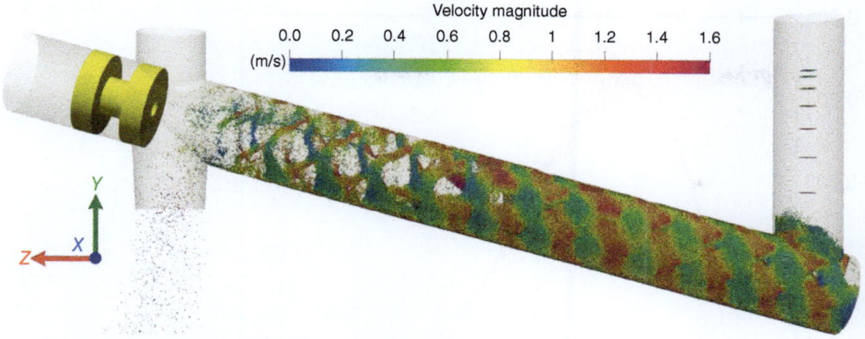

Figure 7.8 DEM modelling of the continuous blending process.

DEM simulations of continuous blending in a full-scale blender have only become feasible lately thanks to the advances in the high-performance computing, especially using GPUs. Figure 7.8 shows a GPU-enhanced DEM simulation of continuous blending in a full-scale continuous blender. The DEM analysis can provide detailed information on continuous blending, such as powder profiles, and in the blender. It can also be used to investigate the effect of the operation parameters and equipment design on the blending performance, e.g. when the steady-state can be reached and how the hold-up material in the blender changes. The hold-up mass is the material mass distributed in the blender when it operates at the steady state. For example, Figure 7.9 shows a typical time evolution of hold-up mass in the blender obtained using the GPU-enhanced DEM. In the figure, the black dash line labels the steady-state period. The information of hold-up mass is vital for the process engineers and can guide blender design and optimisation of processing parameters.

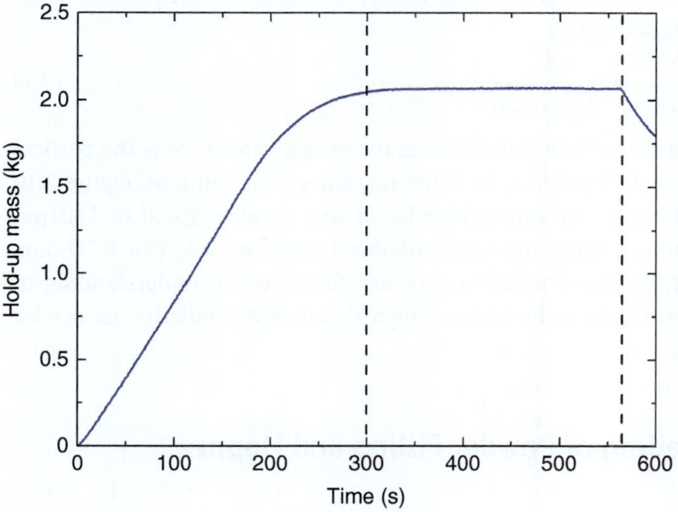

Figure 7.9 Evolution of hold-up mass in the blender.

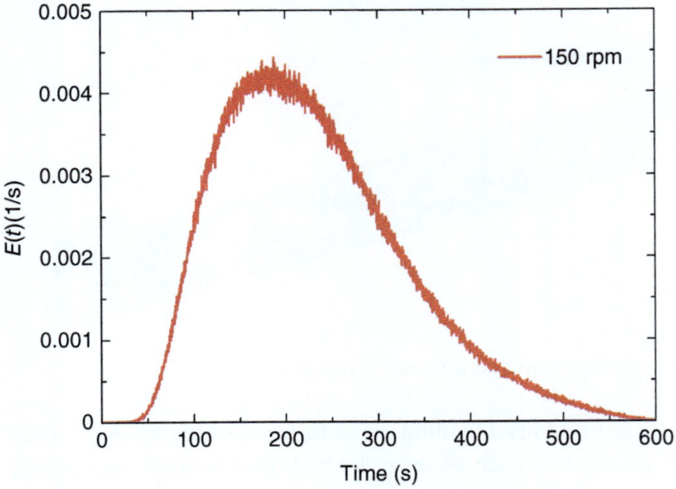

Figure 7.10 Residence time distribution for a continuous blender.

During the blending process, the rotation of the impeller contributes to the conveying and mixing of powders in the blender, which consumes power. Knowing how long the powder stays inside the blender and how much power the process consumes is practically important. RTD is the most effective way to characterise how long the powder resides inside the blender. Experimental approaches can measure the tracing material at the inlet and outlet to determine the residence time while requiring expensive equipment and materials. Moreover, the selection of tracing material is challenging due to the material properties can result in different RTD responses [53]. The tracer material properties are important to identify the RTD accurately and develop an efficient RTD model. By contrast, the DEM approach can readily obtain the residence time [54]. During DEM simulation, the RTD $E(t)$ of the particle in the continuous blender is defined as

$$E(t) = \frac{N(t)}{\int_0^{\infty} N(t)\mathrm{d}t} \cong \frac{N_i}{\sum_{i=0}^{\infty} N_i \Delta t_i} \tag{7.7}$$

where $N(t)$ is the number of particles crossing the exist at time t, Ni is the particle number in the residence time bin i, Δt is the residence time bin size. Figure 7.10 shows a typical RTD in a continuous blender at the impeller speed of 150 rpm (revolutions per minute), using the GPU-enhanced DEM model. The RTD data extracted from the full-scale simulation provides fundamental understanding of the process and useful data to build a detailed data-based predictive model for continuous blending.

7.5 DEM Modelling of Powder Filling and Hopper Discharging

Powder filling and discharging processes are very common in the pharmaceutical industry. Understanding the powder flow behaviour is the key element in designing efficient filling and discharging equipment. Particle attributes and operating conditions play vital roles in defining the flow properties of pharmaceutical powders. For example, Mukherjee et al. [55] examined the effect of humidity on the flowability of pharmaceutical powders. They used DEM to model the discharge of powder of various cohesivity levels from different hopper geometries. Their prediction of the powder flow was in good agreement with the experiments showing the high capability of the DEM technique in predicting the flow behaviour of pharmaceutical powders in different humidity conditions. Hesse et al. [56] conducted DEM simulations and experimental tests of the discharge procedure for a wedge-shaped silo having varying configurations. To incorporate the effect of the fluid, surrounding the powder, on its flow behaviour, they used coupled CFD–DEM simulations, and the results showed low sensitivity of the discharging mass flow rate to the hopper angle variation, and the airflow in the hopper would extend to stagnant zones, resulting in reduced air impairment for flat hopper angles. It demonstrates that DEM simulations can facilitate the design and optimisation of the hopper discharge process in the pharmaceutical industry.

Moreover, pharmaceutical tablets are produced by compressing dry powder blends in a die using a rotary press which consists of three distinct processes: die filling, compaction, and tablet ejection. First, the pharmaceutical powder is filled into a specific die, which is termed the die filling process. The powder in the die is then compacted by a punch to form a tablet. Finally, the tablet is ejected and collected.

Die filling is the process of delivering powder into a die which is essentially a combination of powder discharging and packing processes [57]. It has been pointed out that the powder properties, system configuration, and operation parameters are the main factors, which directly affect the die filling performance and final tablet properties. Die filling is mainly carried out taking the advantage of the gravitational force, i.e. the powder is discharged into a die due to its weight. This type of die filling, which is known as the gravity filling, can be generally performed by moving a powder container (feed shoe) having a downward opening (orifice) on top of a die [58]. The feed shoe may be in form of a cylinder, cone, or other possible shapes, and its orifice can take various shapes and sizes [59]. The die filling performance depends on many parameters, including the feeder type, the operational conditions, and powder properties [60].

During the research and development stage in the pharmaceutical industry, compaction simulators with a simplified linear die filling system are normally used. This system makes it feasible to investigate the powder properties and operation parameters on the product quality, requiring small amount of testing materials. The linear die filling process is widely investigated by many researchers [61, 62]. Figure 7.11 presents the powder flow in a linear die filling rig, which mainly consists of a feeder, a die, and a die table. During the die filling process, the feeder linearly moves on the die table. The powder enters the die when the feeder passed over the die, and packs inside the die.

Figure 7.12 shows the typical variation of the filling ratio with the filling velocity obtained from DEM modelling. It is clear that at a low filling velocity, a high filling ratio is obtained. In contrast, the filling ratio decreased with a further increase in filling velocity. For filling velocity less than around 150 mm/s, the filling ratio can

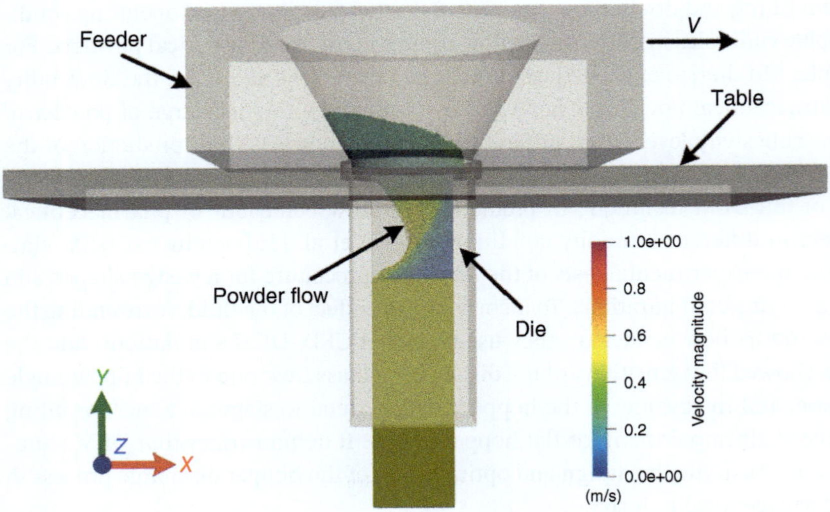

Figure 7.11 DEM modelling of linear die filling.

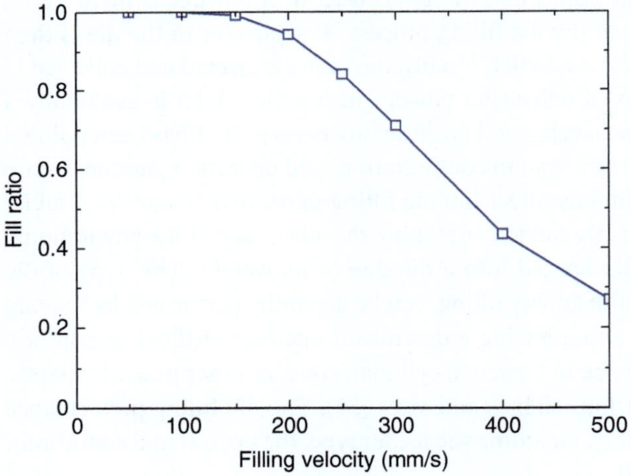

Figure 7.12 Variation of the filling ratio with the filling velocity.

be considered as one that indicates a full die. Thus, there was a maximum filling velocity, which allowed the die to be completely filled. The maximum filling velocity was defined as the critical filling velocity. It could be concluded that when the filling velocity was less than critical filling velocity, the filling ratio was close to unity. The filling ratio will be less than unity when the filling velocity is higher than the critical filling velocity, which demonstrates the process condition can significantly affect the filling behaviour. It is also recognised that the critical filling velocity also depends upon the system design and material properties [63].

In addition to investigating the impact of material properties and process conditions on die filling behaviour, DEM can also be used to guide the equipment design

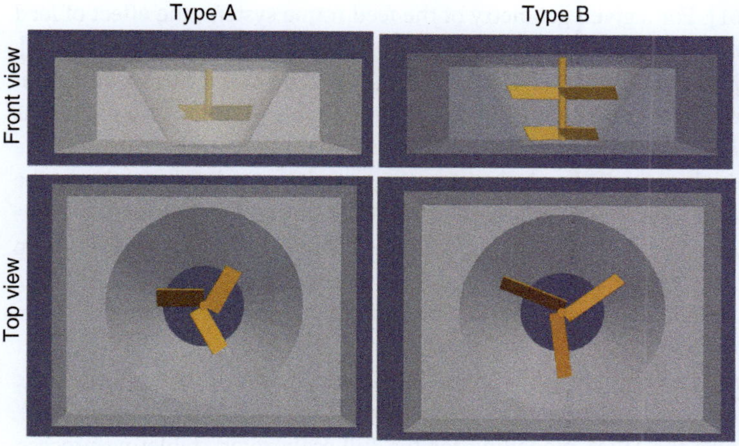

Figure 7.13 A shoe feeder assembled with different stirrer designs.

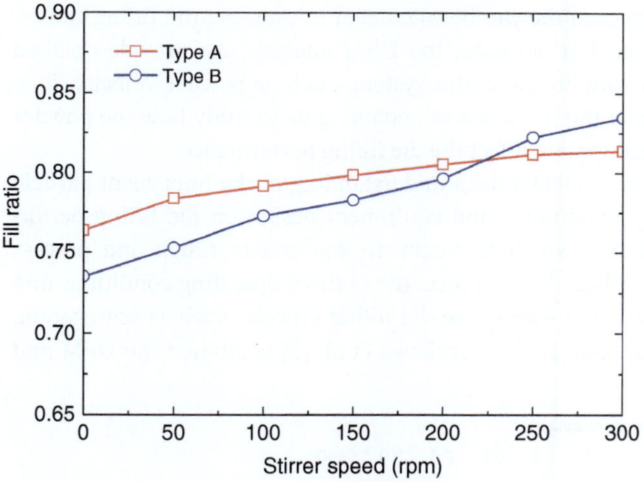

Figure 7.14 Effect of stirrer speed on the fill ratio for different stirrer designs.

and optimisation. For example, a stirrer is generally used in the feed shoe to improve the die filling performance. When the stirrer rotates, the powder is agitated to facilitate the flow or is pushed into the die. For this purpose, the design of the stirrer is critical. Using DEM, the performance of various stirrer designs can be evaluated as illustrated in Figure 7.13, in which Type A is a single-layer triple-blade design, and type B is the two-layers triple-blade design. Filling ratios for various stirrer designs were obtained using DEM and presented in Figure 7.14, in which the effect of stirrer speed on the fill ratio for different stirrer designs under the filling velocity of 300 mm/s is also explored. It can be seen that Type A leads to a more robust filling performance than Type B.

In actual pharmaceutical production, a rotary tablet press equipped with a feed frame is used, in which a stirrer or a paddle is designed to achieve better feeding performance [61]. For a given geometry of the feed frame system, the effect of feed frame configuration on die filling performance was extensively investigated [64]. Additionally, the paddle design and the rotation speed are two critical factors that need to be carefully considered during the design process [65]. Therefore, there is an emerging demand to explore the feeding mechanisms related to the feed frame designs. DEM is then applied to investigate the complex powder dynamics in the feed frame and to determine the influence of feed frame design on the process performance [66]. By employing DEM simulation, RTD inside the feed frame system can be obtained, so can be the tablet mass and mass variability. A good understanding of powder flow inside the freed frame is important in guiding the geometric design and selecting operation parameters.

Recent advances in computing power and DEM make it possible to run large-scale DEM simulations of the actual production process in a reasonable time frame. Figure 7.15 shows the DEM modelling for the full-scale rotary tableting process. Numerical analysis can reveal the powder flow behaviour in the feed frame, which is extremely difficult to achieve with traditional experimental measurements. In addition, the fill mass in each die can be calculated to examine the filling consistency and mass variation. Furthermore, the DEM analysis can provide detailed information on powder flow in the entire system, such as powder velocity, flow pattern, and segregation at the particle level, enabling us to study how the powder properties and process equipment affect the die filling performance.

These DEM simulations enable a deep understanding of the impacts of particle attributes, the operating conditions, and equipment design on die filling performance and product quality, which is crucial in equipment design and process optimisation for the die filling. For instance, the optimal operating conditions and possible issues that may occur during the die filling process, such as segregation, are extensively explored using DEM. Furukawa et al. [5] combined the DEM and

Figure 7.15 DEM modelling of the rotary tableting process.

the response surface methodology (RSM) to investigate the segregation tendency of binary mixtures during die filling in a pharmaceutical tableting machine. They performed a comprehensive calibration study for obtaining reliable model input parameters followed by validation of their simulation results. Their results showed that segregation was significantly affected by particle size, while the die velocity did not show any considerable impact. The prevalent mechanism of segregation was found to be the percolation of the fine particles in the feed shoe and the back corners of the die. Hildebrandt et al. [67] used DEM to assess the powder flow behaviour in a rotary tablet press. Their tablet press had three rotating paddle wheels and a force feeder. They established a direct relationship between the flow of powder in the feeder and the quality of tablets. DEM simulations revealed that the segregation could happen in the force feeder and as a result of shear forces exerted by the paddles. They also suggested some modifications in the design of the force feeder, arrangement of the paddle wheels, and shape of the blades that could diminish the segregation tendency of the powder mixture. DEM modelling has clearly resulted in a better understanding of the filling process and subsequently the quality of pharmaceutical products and patient's safety.

7.6 DEM Modelling of Twin Screw Granulation

Tablets are the most common oral dosage form for pharmaceutical drug delivery, and the formulation generally consists of APIs, and excipients with different functionalities such as fillers, disintegrants, and binders. The constituent ingredients are generally made of small particles (in microns up to hundreds of microns) of different material attributes, which are prone to many manufacturing problems, such as poor flowability and un-wanted agglomeration induced by strong cohesion between fine particles and segregation due to the difference in size, density and other material attributes. Consequently, the formulated powder blends are granulated to improve their flow characteristics and manufacturability. There is an increasing interest in adopting continuous granulation, in particular, twin screw granulation (TSG), to improve manufacturing efficiency and product quality in the pharmaceutical industry. Compared to the batch granulation process (such as high shear granulation), continuous granulation has many advantages, including flexibility in process control and equipment design, good reproducibility, short residence times, small liquid/solid ratios (hence low energy requirement and reduced manufacturing cost) and the ability to granulate difficult-to-process formulations [68].

The twin screw granulator mainly consists of a self-wiping co-rotating intermeshing twin screw, barrel, and feeder, as shown in Figure 7.16. During TSG, the material is added through the inlet and then kneaded and agglomerated in the barrel without increasing pressure. The screw in twin screw granulators is configured using various combinations of the conveying element, the kneading element, and the distributive mixing element (DME). The assembly of different types of screw elements is very flexible due to the module design concept. It is shown that the screw configurations can significantly affect the granulation process [69]. In the past, the design

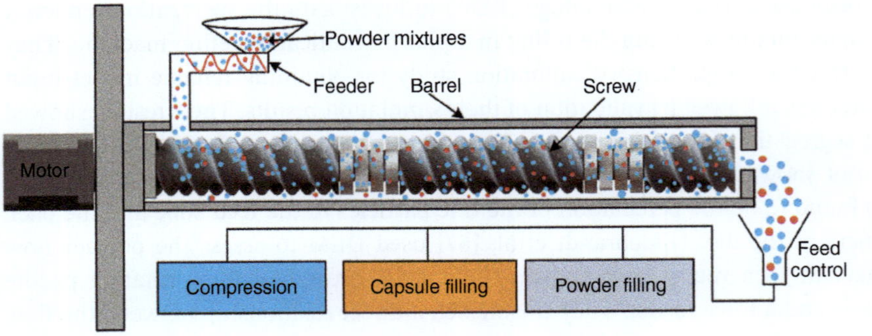

Figure 7.16 Schematic diagram of twin screw granulation.

of twin screw granulators and the assessment of their performance require intensive experimental tests, which are very time-consuming and costly. With numerical techniques, the study and performance evaluation of new granulators have reached a digital era [25].

DEM is widely used to model the granulation process and help the design process. For TSG, residence time and granule properties (i.e. strength and size distribution) can be captured [70]. DEM simulations can be used to evaluate the impact of screw type and screw configuration, thus providing detailed performance characteristics and guiding the equipment design. Figure 7.17 shows the twin screw granulators made of different element types, including long pitch conveying elements (LCE), short pitch conveying elements (SCE), kneading elements with a 30° staggering angle (KE30°), kneading elements with a 60° staggering angle (KE60°), kneading elements with a 90° staggering angle (KE90°), and the DME.

Figure 7.18 presents the corresponding RTDs of particles in the abovementioned twin screw granulators with various element types (Figure 7.16). It is clear that the screw element type strongly impacts the RTD, which provides useful data for the process design. It can be found that the RTDs for LCEs and SCEs are much narrow than that of the other element type during TSG. According to different requirements of products, screw element types can be determined based on the numerical results. Therefore, DEM modelling can provide detailed information at the particle-based level, which helps further understand the TSG process. It can also be used to evaluate the performance of a given design of twin screw granulators.

The powder properties such as granule size and granule shape strongly affect the conveying characteristics of powder in the granulators. Particles with different shapes contribute to different bulk responses during the granulation, finally leading to different power consumption, as reported by Zheng et al. [70]. Power consumption is a key indicator that needs to be considered during the geometric design and process design. DEM numerical results can be used to assist decision-making in the formulation and process design, thus significantly reducing development time and cost.

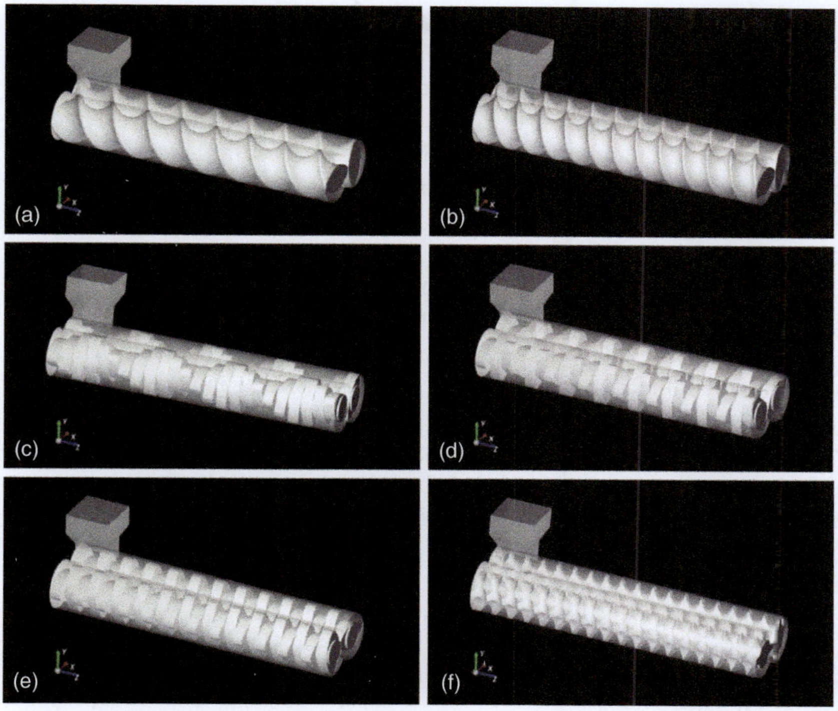

Figure 7.17 Model twin screw granulators with various screw elements: (a) long pitch conveying elements; (b) Short pitch conveying elements; (c) Kneading elements with a 30° staggering angle; (d) Kneading elements with a 60° staggering angle; (e) Kneading elements with a 90° staggering angle; (f) Distributive mixing elements.

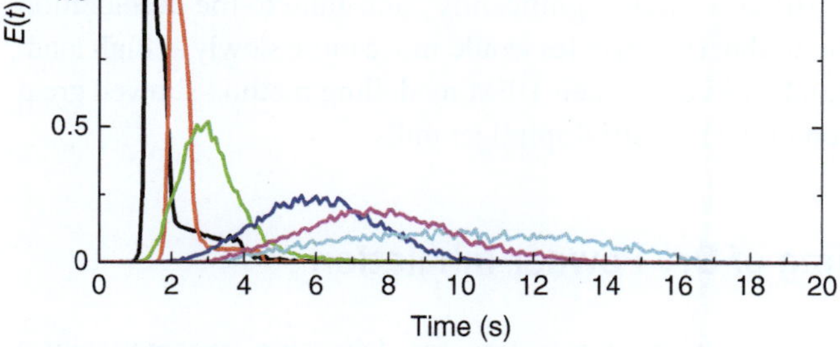

Figure 7.18 Residence time distributions obtained with twin screw granulators with various screw elements shown in Figure 7.16.

7.7 DEM Modelling of Milling

Milling is an important industrial process that uses mechanical means for reducing the size or changing the shape of particulate materials such as minerals, chemicals, pharmaceuticals, and agricultural products. In the pharmaceutical industry, milling is also performed in the amorphisation of drugs and to enhance their flowability and solubility [71]. A variety of mills, such as pin mill, hammer mill, jet mill, and ball mill are used in processing pharmaceutical powders, which reduce particle sizes through mechanisms of impact, attrition, compression, or a combination of them. The pin and hammer mills operate based on the impact mechanism, and the jet mill works based on attrition. Ball mills, including the single ball mill, planetary ball mill, and tumbling ball mill, function based on a combination of the impact and attrition mechanisms, while in the conical screen mills, compression is responsible for size reduction. Despite being widely used in the pharmaceutical industry, milling is still one of the least understood unit operations because of the complex combination of process and material variables as well as milling mechanisms [72]. One of the great challenges in the milling process is its considerably low energy efficiency. A better understanding of the milling mechanisms is the key to optimisation of the milling process and improving the energy utilisation efficiency.

DEM is widely regarded as a useful tool to explore the underlying mechanisms of the milling process that otherwise would be hardly understood [73, 74]. To analyse the flow and breakage of the particulate materials in the mills, for example, Metzger and Glasser [75] simulated the breakage of the bonded agglomerates in a batch ball mill using DEM with the bonded particle model (BPM). They found that it is possible to tune the operating conditions in connection with the mechanical properties of the materials to improve the performance of the ball mills [75]. Bonakdar and Ghadiri [76] established a relationship for the prediction of the particle's breakage in pin mills. In some mills, like the jet mills, milling occurs as a result of the interaction of the solid and fluid phases. In such cases, the DEM is coupled with the computational fluid dynamics to model the effects of both solid and fluid phases, respectively. For instance, Bhonsale et al. [77] examined the dynamics of particles in a spiral jet mill using a coupled CFD–DEM model, and the effects of operating pressure and powder loading on particles velocity at different positions of the mill. They observed that the particles would significantly contribute to the deceleration of the fluid. They also found that the particles would move more slowly at high loadings and low pressures and vice versa. Their DEM modelling method showed great potential for optimisation of the industrial spiral jet mills.

7.8 DEM Modelling of Dry Powder Inhalation

Dry powder inhalation is a noninvasive system for delivering systematically a wide range of drug powders from small molecules to very large proteins, such as insulin [78]. DPIs are hence increasingly used to deliver medicines into lungs and airways for rapid and efficient treatment of these diseases, such as chronic obstructive pulmonary disease (COPD), asthma, and other respiratory diseases, while minimising the dose requirement as they can achieve a similar or superior therapeutic effect with part of the systemic dose. The performance and efficiency of DPIs are dictated by three processes: agglomeration, de-agglomeration, and deposition [79]. Thus far, *in vitro* tests using impactors are routinely performed to assess such performance of DPIs [80]. However, *in vitro* tests are mainly based on the trial-and-error approach, which is not only very costly and time-consuming, but also unreliable to demonstrate bioequivalence of inhalation products. Hence, numerical analyses of dry powder inhalation using DEM or DEM–CFD have attracted increasing interest in DPI [81–83], in particular on modelling of agglomeration, de-agglomeration, and deposition processes.

Drug particles used in DPIs are normally less than 5 μm in diameter and bound to be cohesive and have a poor flowability. To overcome such poor flowability and improve aerosolization performance of DPIs, drug particles are either attacked to coarse carrier particles (usually ~70 μm) to form carrier-based agglomerates or aggregated into drug-only agglomerates [81]. This is the so-called agglomeration process, which is always conducted in a blender [84], and is crucial for controlling the efficiency of DPIs. During blending, it is expected that the fine drug particles are sticking to the surfaces of carriers due to the strong cohesive forces, e.g. electrostatic forces, van der Waals (VDW) forces, or a combination of both [85]. Using DEM, Yang et al. [81] simulated the agglomeration process by mixing drug particles and a single carrier particle in a vertically vibration container, as shown in Figure 7.19. The adhesion force that governed the agglomeration process was VDM forces. It was found that an optimal frequency and amplitude for the vibration velocity were obtained which can reach the maximum number of drug particles sticking with the carrier. Yang et al. [83] also explored the agglomeration process dominated by electrostatic attractive interactions using DEM, in which drug particles and carrier particles were initially charged with positively and negatively, respectively. The effect of vibrational velocity amplitude and frequency for the vibration container were also examined. It was found the contact number decreases with increasing frequency and velocity amplitude of the vibration, which is different from that dominated by VDW interactions reported in ref. [81], as shown in Figure 7.20.

Ideally, the formed agglomerates for DPIs should be strong enough during manufacturing but weak enough to be disintegrated and dispersed during inhalation, i.e.

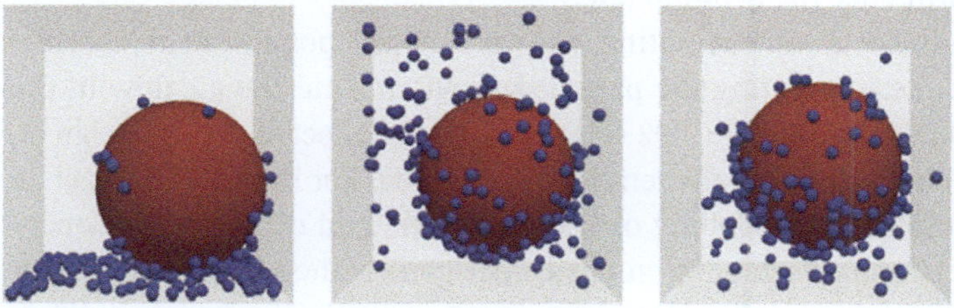
(a) $t = 0.0750$ s (b) $t = 0.1205$ s (c) $t = 0.2251$ s

Figure 7.19 Snapshots of carrier and drug particles during vibration at different time instances. Source: Yang et al. [81]/ with permission of Springer.

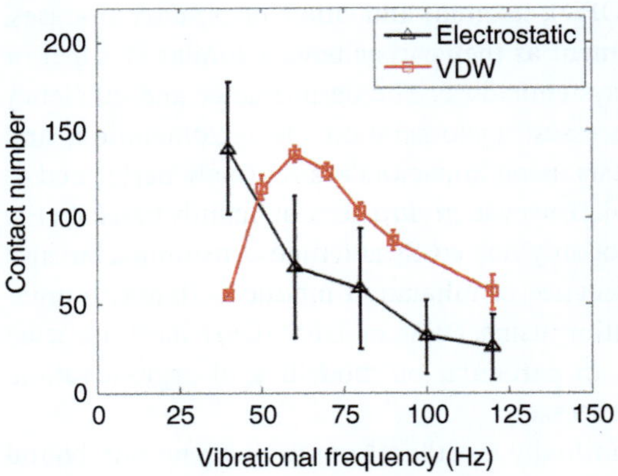

Figure 7.20 Comparison of the contact number during mixing process governed by electrostatic force and van der Waals force (Vibrational amplitude = 0.04 m/s, Electrostatic: charge on the carrier = 0.042 nC, VDW: surface energy = 0.006 J/m^2). Source: Reproduced from Yang et al. [83]/with permission of Elsevier.

during de-agglomeration [86]. The de-agglomeration process in DPIs involves air flow-induced aerodynamic forces and agglomeration-device impactions [79, 87]. The aerosolization performance of DPI is the combined effects of air aerodynamic forces, mechanical impactions, and particle-particle interactions [86]. Using DEM-CFD, de-agglomeration for both drug-only and carrier-based agglomerates was modelled [88–91]. Tong et al. [89] simulated the dispersion process of drug mannitol agglomerates in a cyclonic flow at various gas velocities where the agglomerate was formed by polydisperse particles. They found the flow rate significantly influences the dispersion behaviour. For the agglomerate with finer primary particles, it is difficult to disperse at a relatively lower flow rate due to the large cohesion but turns to have a higher dispersion efficiency at a relatively higher flow rate. Furthermore, they pointed out that the air flow would also cause agglomerate-wall impact. Calvert et al. [88, 90] also modelled the de-agglomeration of a drug-only agglomerate with different relative velocities between particles and the air. It is found the agglomerate disintegrates rapidly, i.e. interparticle bonds break, when the relative velocity reach a threshold value and the threshold velocity decreased with the size of the agglomerate. The interplay of both turbulence and adhesion force on the de-agglomeration of a 'clump' of fine particles in homogeneous isotropic turbulence were explored in an Eulerian–Lagrangian framework by Yao et al. [91]. They found the agglomerate size, Reynolds number (refer to the air velocity), and adhesion number (refer to the work of adhesion) contribute to the breakup of the agglomerate and influence the breakup rate.

Recent advances in DEM and computing hardware made it possible to model DPIs at the device-scale, i.e. simulating the particle bahaviours and the air flow in the whole commercial DPI products, [82, 92–94], enabling the effect of device design on the aerosolisation and deposition efficiency to be explored. For instance, Zhou et al. [92] carried out DEM-CFD modelling of DPI Aerolizer® and explored the dependence of the dispersion performance on device designs. In the simulation, the air inlet size, mouthpiece length, and grid voidage were systematically altered, and they showed that the air inlet size and grid structure of DPI played important roles in the dispersion performance. Benque and Khinast [93] also stimulated the air flow field and the trajectory of carrier particle in an Aerolizer inhaler without any fine drug particles but considered three typical different inhalation air rates from adults, children, and COPD patients. They found that there is a larger velocity fluctuation for carrier particles with inhalation profiles from patients than that with the constant air rate flow. Alfano et al. [94] conducted a two-stage DEM–CFD simulation on the swirling flow in a simplified NEXThaler® inhaler. In the first stage, only the carrier particles were modelled to obtain the trajectories and velocity of the carrier particles. In the second stage, drug particles were mixed with carrier particles with adhesive force and the de-agglomeration process in the DPI was simulated. Numerical simulations considered both carrier and drug particles were also performed to investigate the emitted aerodynamic particle size distribution and drug delivery efficiency from DPI to a tracheobronchial tree by Zhao et al. [82], in which different inhalation flow rates and DPI designs (Spiriva® Handihaler® DPI and a generic DPI) were compared. In addition, the particle shape of the carriers with various aspect ratios was also considered to examine its effects on particle aerodynamics. They found the drug delivery efficiency decreased with the inhalation flow rates from 30 to 60 L/m and then increased sharply when the inhalation flow rates reached 90 L/m. The shape of lactose carriers was found to affect the drug delivery efficiency significantly, but it had little influence on the lung deposition patterns of the drug particles.

7.9 Summary and Perspectives

The DEM is a numerical approach that models powder as assemblies of individual particles, in which the attributes (i.e., physical and material properties) of individual particles are explicitly considered and their dynamics are obtained. Based upon the detailed information on individual particles, the bulk behaviour of powders in complex pharmaceutical processing can be determined. DEM is also successfully coupled with some fluid solvers, in particular, CFD, so that complex solid–fluid flows can be modelled. Although the early applications of DEM focused on understanding the underlying mechanisms in various pharmaceutical processes, recent advances in DEM and high-performance computing enable simulations of large particle systems in full-scale pharmaceutical processing. Consequently, it is now possible to use DEM for equipment design and process optimisation. The digital models based upon DEM pave a solid foundation for developing robust digital twins for a wide range of pharmaceutical processes, in which the digital models could be robustly paired with the physical equipment taking full advantage of the emerging internet of things and advanced sensor technologies. This will make a paradigm shift in equipment design and pharmaceutical manufacturing.

Nomenclature

C_r coefficient of rolling friction
CR_{Ae} contact ratios of API-excipient
CR_{AA} contact ratios of API–API

d	distance
$E(t)$	residence time distribution function
F_c	contact force
F_{nc}	non-contact force
g	gravity acceleration
H	height
i	residence time bin number
I	moment of inertia
M	moment
m	mass
$N(t)$	the number of particles crossing that exists at time t
N_A	number of API particles
N_e	number of excipient particles
T_R	Rayleigh time step
t	time
R	radius
V	velocity
X	position
ν	Poisson's ratio
ρ	density
τ	Shear modulus
ω	angular velocity

References

1 UN Sustainable Development Goals United Nations: Department of Economic and Social Affairs (2022). https://sdgs.un.org/goals.

2 Cundall, P.A. and Strack, O.D.L. (1979). A discrete numerical model for granular assemblies. *Géotechnique* 29 (1): 47–65.

3 Nakamura, H., Iwasaki, T., and Watano, S. (2006). Numerical simulation of film coating process in a novel rotating fluidized bed. *Chemical and Pharmaceutical Bulletin* 54 (6): 839–846.

4 Fu, X., Dutt, M., Bentham, A.C. et al. (2006). Investigation of particle packing in model pharmaceutical powders using X-ray microtomography and discrete element method. *Powder Technology* 167 (3): 134–140.

5 Furukawa, R., Shiosaka, Y., Kadota, K. et al. (2016). Size-induced segregation during pharmaceutical particle die filling assessed by response surface methodology using discrete element method. *Journal of Drug Delivery Science and Technology* 35: 284–293.

6 Hildebrandt, C., Gopireddy, S.R., Scherließ, R., and Urbanetz, N.A. (2018). Simulation of particle size segregation in a pharmaceutical tablet press lab-scale gravity feeder. *Advanced Powder Technology* 29 (3): 765–780.

7 Yeom, S.B. and Choi, D.H. (2019). Scale-up strategy in quality by design approach for pharmaceutical blending process with discrete element method simulation. *Pharmaceutics* 11 (6): 264.

8 Hassanpour, A., Pasha, M., and Alizadeh, B.M. (2014). Discrete element method applications in process engineering. In: *Introduction to Software for Chemical Engineers* (ed. M.M. Martin), 245–286. Boca Raton: CRC Press.

9 Ning, Z. and Ghadiri, M. (2006). Distinct element analysis of attrition of granular solids under shear deformation. *Chemical Engineering Science* 61 (18): 5991–6001.

10 Hertz, H. (1882). Ueber die Berührung fester elastischer Körper (on the contact of elastic solids). *Journal für die reine und angewandte Mathematik (Crelle's Journal)* 92: 156–171.

11 Deresiewicz H, Mindlin RD, Columbia U, Department of Civil E. Elastic spheres in contact under varying oblique forces. University of Columbia 1952.

12 Thornton, C. (2015). *Granular Dynamics, Contact Mechanics and Particle System Simulations.* Springer.

13 Johnson, K.L., Kendall, K., and Roberts, A.D. (1971). Surface energy and the contact of elastic solids. *Proceedings of the Royal Society of London A: Mathematical, Physical and Engineering Sciences* 324 (1558): 301–313.

14 Pasha, M., Dogbe, S., Hare, C. et al. (2014). A linear model of elasto-plastic and adhesive contact deformation. *Granular Matter* 16 (1): 151–162.

15 Mindlin RD, Deresiewicz H. Elastic spheres in contact under varying oblique forces. *ASME Journal of Applied Mechanics,* 1953; 20(3): 327–344, https://doi.org/10.1115/1.4010702.

16 Martín, M.M. (2014). *Introduction to Software for Chemical Engineers.* CRC Press.

17 Seville, J.P. and Wu, C.-Y. (2016). *Particle Technology and Engineering: An Engineer's Guide to Particles and Powders: Fundamentals and Computational Approaches.* Butterworth-Heinemann.

18 Marigo, M. and Stitt, E.H. (2015). Discrete element method (DEM) for industrial applications: comments on calibration and validation for the modelling of cylindrical pellets. *KONA Powder and Particle Journal* 32: 236–252.

19 Alizadeh, M., Asachi, M., Ghadiri, M. et al. (2018). A methodology for calibration of DEM input parameters in simulation of segregation of powder mixtures, a special focus on adhesion. *Powder Technology* 339: 789–800.

20 Coetzee, C.J. (2017). Review: calibration of the discrete element method. *Powder Technology* 310: 104–142.

21 Boikov, A.V., Savelev, R.V., and Payor, V.A. (ed.) (2018). DEM calibration approach: random forest. In: *Journal of Physics: Conference Series.* IOP Publishing.

22 El Kassem, B., Salloum, N., Brinz, T. et al. (2021). A semi-automated DEM parameter calibration technique of powders based on different bulk responses extracted from Auger Dosing experiments. *KONA Powder and Particle Journal* 38: 235–250.

23 Huang, A.-N. and Kuo, H.-P. (2014). Developments in the tools for the investigation of mixing in particulate systems – A review. *Advanced Powder Technology* 25 (1): 163–173.

24 Chaudhuri, B., Mehrotra, A., Muzzio, F.J., and Tomassone, M.S. (2006). Cohesive effects in powder mixing in a tumbling blender. *Powder Technology* 165 (2): 105–114.

25 Behjani, M.A., Motlagh, Y.G., Bayly, A.E., and Hassanpour, A. (2020). Assessment of blending performance of pharmaceutical powder mixtures in a continuous mixer using discrete element method (DEM). *Powder Technology* 366: 73–81.

26 Ahmadian, H., Hassanpour, A., and Ghadiri, M. (2011). Analysis of granule breakage in a rotary mixing drum: experimental study and distinct element analysis. *Powder Technology* 210 (2): 175–180.

27 Hassanpour, A., Tan, H., Bayly, A. et al. (2011). Analysis of particle motion in a paddle mixer using discrete element method (DEM). *Powder Technology* 206 (1–2): 189–194.

28 Lemieux, M., Bertrand, F., Chaouki, J., and Gosselin, P. (2007). Comparative study of the mixing of free-flowing particles in a V-blender and a bin-blender. *Chemical Engineering Science* 62 (6): 1783–1802.

29 Lemieux, M., Léonard, G., Doucet, J. et al. (2008). Large-scale numerical investigation of solids mixing in a V-blender using the discrete element method. *Powder Technology* 181 (2): 205–216.

30 He, Y., Bayly, A.E., Hassanpour, A. et al. (2020). Flow behaviour of an agitated tubular reactor using a novel dynamic mesh based CFD model. *Chemical Engineering Science* 212: 115333.

31 Arratia, P., Duong, N.-h., Muzzio, F. et al. (2006). A study of the mixing and segregation mechanisms in the Bohle Tote blender via DEM simulations. *Powder Technology* 164 (1): 50–57.

32 Basinskas, G. and Sakai, M. (2016). Numerical study of the mixing efficiency of a ribbon mixer using the discrete element method. *Powder Technology* 287: 380–394.

33 Alian, M., Ein-Mozaffari, F., Upreti, S.R., and Wu, J. (2015). Using discrete element method to analyze the mixing of the solid particles in a slant cone mixer. *Chemical Engineering Research and Design* 93: 318–329.

34 Hogg, R. (2009). Mixing and segregation in powders: evaluation, mechanisms and processes. *KONA Powder and Particle Journal* 27: 3–17.

35 Brone, D. and Muzzio, F.J. (1997). Size segregation in vibrated granular systems: a reversible process. *Physical Review E* 56 (1): 1059–1063.

36 Gillemot, K.A., Somfai, E., and Borzsonyi, T. (2017). Shear-driven segregation of dry granular materials with different friction coefficients. *Soft Matter* 13 (2): 415–420.

37 Yu, S., Guo, Y., and Wu, C.Y. (2009). DEM/CFD modelling of the deposition of dilute granular systems in a vertical container. *Chinese Science Bulletin* 54 (23): 4318–4326.

38 Hajra, S.K., Shi, D., and McCarthy, J.J. (2012). Granular mixing and segregation in zigzag chute flow. *Physical Review E* 86 (6): 061318.

39 Wangchai, S., Hastie, D.B., and Wypych, P.W. (2018). Particle size segregation of bulk material in dustiness testers via DEM simulation. *Particulate Science and Technology* 36 (1): 20–28.

40 Guo, Y., Wu, C.-Y., Kafui, K., and Thornton, C. (2011). 3D DEM/CFD analysis of size-induced segregation during die filling. *Powder Technology* 206 (1–2): 177–188.

41 Pasha, M., Hare, C., Ghadiri, M. et al. (2016). Effect of particle shape on flow in discrete element method simulation of a rotary batch seed coater. *Powder Technology* 296: 29–36.

42 Alizadeh, M., Hassanpour, A., Pasha, M. et al. (2017). The effect of particle shape on predicted segregation in binary powder mixtures. *Powder Technology* 319: 313–322.

43 Asachi, M., Behjani, M.A., Nourafkan, E., and Hassanpour, A. (2021). Tailoring particle shape for enhancing the homogeneity of powder mixtures: experimental study and DEM modelling. *Particuology* 54: 58–68.

44 Drahun, J. and Bridgwater, J. (1983). The mechanisms of free surface segregation. *Powder Technology* 36 (1): 39–53.

45 Zeilstra, C., Van Der Hoef, M., and Kuipers, J. (2008). Simulation of density segregation in vibrated beds. *Physical Review E* 77 (3): 031309.

46 Guo, Y., Wu, C.-Y., Kafui, K., and Thornton, C. (2010). Numerical analysis of density-induced segregation during die filling. *Powder Technology* 197 (1–2): 111–119.

47 Kleinebudde, P., Khinast, J., and Rantanen, J. (2017). *Continuous Manufacturing of Pharmaceuticals*. Wiley.

48 Pernenkil, L. and Cooney, C.L. (2006). A review on the continuous blending of powders. *Chemical Engineering Science* 61 (2): 720–742.

49 Pauli, V., Roggo, Y., Pellegatti, L. et al. (2019). Process analytical technology for continuous manufacturing tableting processing: a case study. *Journal of Pharmaceutical and Biomedical Analysis* 162: 101–111.

50 Gao, Y., Muzzio, F.J., and Ierapetritou, M.G. (2012). Optimizing continuous powder mixing processes using periodic section modeling. *Chemical Engineering Science* 80: 70–80.

51 Sarkar, A. and Wassgren, C.R. (2012). Comparison of flow microdynamics for a continuous granular mixer with predictions from periodic slice DEM simulations. *Powder Technology* 221: 325–336.

52 Jourdan, N., Neveux, T., Potier, O. et al. (2019). Compartmental modelling in chemical engineering: a critical review. *Chemical Engineering Science* 210: 115196.

53 Escotet-Espinoza, M.S., Moghtadernejad, S., Oka, S. et al. (2019). Effect of tracer material properties on the residence time distribution (RTD) of continuous powder blending operations. Part I of II: experimental evaluation. *Powder Technology* 342: 744–763.

54 Dubey, A., Sarkar, A., Ierapetritou, M. et al. (2011). Computational approaches for studying the granular dynamics of continuous blending processes, 1–DEM based methods. *Macromolecular Materials and Engineering* 296 (3–4): 290–307.

55 Mukherjee, R., Mao, C., Chattoraj, S., and Chaudhuri, B. (2018). DEM based computational model to predict moisture induced cohesion in pharmaceutical powders. *International Journal of Pharmaceutics* 536 (1): 301–309.

56 Hesse, R., Krull, F., and Antonyuk, S. (2020). Experimentally calibrated CFD-DEM study of air impairment during powder discharge for varying hopper configurations. *Powder Technology* 372: 404–419.

57 Xie, X. and Puri, V.M. (2006). Uniformity of powder die filling using a feed shoe: a review. *Particulate Science and Technology* 24 (4): 411–426.

58 Goh, H.P., Heng, P.W.S., and Liew, C.V. (2018). Understanding effects of process parameters and forced feeding on die filling. *European Journal of Pharmaceutical Sciences* 122: 105–115.

59 Mendez, R., Muzzio, F., and Velazquez, C. (2010). Study of the effects of feed frames on powder blend properties during the filling of tablet press dies. *Powder Technology* 200 (3): 105–116.

60 Mehrotra, A., Chaudhuri, B., Faqih, A. et al. (2009). A modeling approach for understanding effects of powder flow properties on tablet weight variability. *Powder Technology* 188 (3): 295–300.

61 Peeters, E., De Beer, T., Vervaet, C., and Remon, J.-P. (2015). Reduction of tablet weight variability by optimizing paddle speed in the forced feeder of a high-speed rotary tablet press. *Drug Development and Industrial Pharmacy* 41 (4): 530–539.

62 Tanimura, S., Singh, R., Román-Ospino, A.D., and Ierapetritou, M. (2021). Residence time distribution modelling and in line monitoring of drug concentration in a tablet press feed frame containing dead zones. *International Journal of Pharmaceutics* 592: 120048.

63 Zhong, W.-Z., Zakhvatayeva, A., Zhang, L., and Wu, C.-Y. (2021). Powder flow during linear and rotary die filling. *International Journal of Pharmaceutics* 602: 120654.

64 Dülle, M., Özcoban, H., and Leopold, C. (2018). Investigations on the residence time distribution of a three-chamber feed frame with special focus on its geometric and parametric setups. *Powder Technology* 331: 276–285.

65 Grymonpré, W., Prudilova, B.B., Vanhoorne, V. et al. (2018). Downscaling of the tableting process: feasibility of miniaturized forced feeders on a high-speed rotary tablet press. *International Journal of Pharmaceutics* 550 (1–2): 477–485.

66 Siegmann, E., Forgber, T., Toson, P. et al. (2020). Powder flow and mixing in different tablet press feed frames. *Advanced Powder Technology* 31 (2): 770–781.

67 Hildebrandt, C., Gopireddy, S.R., Scherließ, R., and Urbanetz, N.A. (2019). Investigation of powder flow within a pharmaceutical tablet press force feeder – A DEM approach. *Powder Technology* 345: 616–632.

68 Suresh, P., Sreedhar, I., Vaidhiswaran, R., and Venugopal, A. (2017). A comprehensive review on process and engineering aspects of pharmaceutical wet granulation. *Chemical Engineering Journal* 328: 785–815.

69 Rahimi, S.K., Paul, S., Sun, C.C., and Zhang, F. (2020). The role of the screw profile on granular structure and mixing efficiency of a high-dose hydrophobic drug formulation during twin screw wet granulation. *International Journal of Pharmaceutics* 575: 118958.

70 Zheng, C., Govender, N., Zhang, L., and Wu, C.-Y. (2022). GPU-enhanced DEM analysis of flow behaviour of irregularly shaped particles in a full-scale twin screw granulator. *Particuology* 61: 30–40.

71 Naik, S. and Chaudhuri, B. (2015). Quantifying dry milling in pharmaceutical processing: a review on experimental and modeling approaches. *Journal of Pharmaceutical Sciences* 104 (8): 2401–2413.

72 Clement, S. and Purutyan, H. (2002). Narrowing down equipment choices for particle-size reduction. *Chemical Engineering Progress* 98 (6): 50–54.

73 Pizette, P., Govender, N., Abriak, N.-E., and Wilke, D.N. (ed.) (2016). GPU DEM simulations and experimental studies of ball milling process for various particle shapes. In: *International Conference on Discrete Element Methods*. Springer.

74 Daraio, D., Villoria, J., Ingram, A. et al. (2020). Using discrete element method (DEM) simulations to reveal the differences in the γ-Al_2O_3 to α-Al_2O_3 mechanically induced phase transformation between a planetary ball mill and an attritor mill. *Minerals Engineering* 155: 106374.

75 Metzger, M.J. and Glasser, B.J. (2013). Simulation of the breakage of bonded agglomerates in a ball mill. *Powder Technology* 237: 286–302.

76 Bonakdar, T. and Ghadiri, M. (2018). Analysis of pin milling of pharmaceutical materials. *International Journal of Pharmaceutics* 552 (1): 394–400.

77 Bhonsale, S., Scott, L., Ghadiri, M., and Van Impe, J. (2021). Numerical simulation of particle dynamics in a spiral jet mill via coupled CFD–DEM. *Pharmaceutics* 13 (7): 937.

78 Wolff, R. (1998). Safety of inhaled proteins for therapeutic use. *Journal of Aerosol Medicine* 11 (4): 197–219.

79 Yang, J., Wu, C.-Y., and Adams, M. (2015). Numerical modelling of agglomeration and deagglomeration in dry powder inhalers: a review. *Current Pharmaceutical Design* 21 (40): 5915–5922.

80 Dorosz, A., Żaczek, M., and Moskal, A. (2021). Dynamics of aerosol generation and release–dry powder inhaler performance considerations. *Journal of Aerosol Science* 151: 105673.

81 Yang, J., Wu, C.-Y., and Adams, M. (2013). DEM analysis of particle adhesion during powder mixing for dry powder inhaler formulation development. *Granular Matter* 15 (4): 417–426.

82 Zhao, J., Haghnegahdar, A., Feng, Y. et al. (2022). Prediction of the carrier shape effect on particle transport, interaction and deposition in two dry powder inhalers and a mouth-to-G13 human respiratory system: a CFD–DEM study. *Journal of Aerosol Science* 160: 105899.

83 Yang, J., Wu, C.-Y., and Adams, M. (2015). DEM analysis of the effect of electrostatic interaction on particle mixing for carrier-based dry powder inhaler formulations. *Particuology* 23: 25–30.

84 Le, V., Thi, T.H., Robins, E., and Flament, M.P. (2012). Dry powder inhalers: study of the parameters influencing adhesion and dispersion of fluticasone propionate. *AAPS PharmSciTech* 13 (2): 477–484.

85 Kaialy, W. (2016). A review of factors affecting electrostatic charging of pharmaceuticals and adhesive mixtures for inhalation. *International Journal of Pharmaceutics* 503 (1–2): 262–276.

86 Zhou, Q.T. and Morton, D.A. (2012). Drug–lactose binding aspects in adhesive mixtures: controlling performance in dry powder inhaler formulations by altering lactose carrier surfaces. *Advanced Drug Delivery Reviews* 64 (3): 275–284.

87 Voss, A. and Finlay, W.H. (2002). Deagglomeration of dry powder pharmaceutical aerosols. *International Journal of Pharmaceutics* 248 (1–2): 39–50.

88 Calvert, G., Hassanpour, A., and Ghadiri, M. (2011). Mechanistic analysis and computer simulation of the aerodynamic dispersion of loose aggregates. *Chemical Engineering Research and Design* 89 (5): 519–525.

89 Tong, Z., Yang, R., Chu, K. et al. (2010). Numerical study of the effects of particle size and polydispersity on the agglomerate dispersion in a cyclonic flow. *Chemical Engineering Journal* 164 (2–3): 432–441.

90 Calvert, G., Hassanpour, A., and Ghadiri, M. (2013). Analysis of aerodynamic dispersion of cohesive clusters. *Chemical Engineering Science* 86: 146–150.

91 Yao, Y. and Capecelatro, J. (2021). Deagglomeration of cohesive particles by turbulence. *Journal of Fluid Mechanics* 911 (A10): 1–26.

92 Zhou, Q.T., Tong, Z., Tang, P. et al. (2013). Effect of device design on the aerosolization of a carrier-based dry powder inhaler—a case study on Aerolizer® Foradile®. *The AAPS Journal* 15 (2): 511–522.

93 Benque, B. and Khinast, J.G. (2021). Estimating inter-patient variability of dispersion in dry powder inhalers using CFD–DEM simulations. *European Journal of Pharmaceutical Sciences* 156: 105574.

94 Alfano, F.O., Benassi, A., Gaspari, R. et al. (2021). Full-scale dem simulation of coupled fluid and dry-coated particle flow in swirl-based dry powder inhalers. *Industrial and Engineering Chemistry Research* 60 (42): 15310–15326.

8

Algorithms and Capabilities of *Solidity* to Simulate Interactions and Packing of Complex Shapes

Jiansheng Xiang and John-Paul Latham

Imperial College London, Department of Earth Science and Engineering, Faculty of Engineering, Prince Consort Road, London SW7 2AZ, UK

Abbreviation

CPU	central processing unit
DEM	discrete element method
FDEM	combined finite-discrete element method
FEM	finite element method
MPI	message passing interface
OpenMP	open multi-processing

8.1 Introduction

Particle packing has been studied for many years as it is of fundamental importance to many industries: mining, chemical engineering, pharmaceuticals, agriculture, food handling, etc. In previous years, most of the published papers on simulating packing of spherical particles were about using discrete element method (DEM) which was invented by Cundall and Strack [1].

Recently, the packing of complex shapes has been attracting more and more interest motivated by their wide application in industry. A number of numerical algorithms for simulation of particle packing have been published. Torquato and Jiao [2] used an adaptive fundamental cell subject to periodic boundary conditions (the 'adaptive shrinking cell' scheme) to obtain densest random packings of Platonic and Archimedean solids. DigiPack [3] was used to simulate the structure pattern of catalyst pellets. In these two studies, particle–particle interaction was not considered. These algorithms are also divided into pure geometric types. To consider the effect of the contact mechanics, some researchers also independently developed what are now termed clustered overlapping sphere algorithms [4–6]

Simulations in Bulk Solids Handling: Applications of DEM and other Methods, First Edition.
Edited by Don McGlinchey.

to represent non-spherical shapes and simulate them in DEM. Cleary [7, 8] used the so-called superquadrics and formulas to define ellipsoids or other quadrics. Greben et al. [9] use a commercial code popular in the video games industry, PhysX (by Navidia) to place packs of armour units such as Antifer Cubes and Dolosse. Blender is a similar module with certain physically realistic features used to introduce mechanical realism into computer animation visualisations. Recently, Polyhedral/Polygon DEM [10–12] in which the particle shape is represented with surface mesh elements are used to simulate rigid or non-spherical shapes. However, most of them can only deal with simple and regular shapes due to the complex and expensive numerical algorithms needed to simulate complex shapes and deformable solid.

Finite-discrete element method (FDEM) combines the finite element method, (FEM), for modelling stress and deformation of solids, and DEM for modelling the collision and motion of solids. Important for handling complex shapes, FDEM solves the contact mechanics using a distributed contact force approach. The parallelised FDEM code 'Solidity' has been developed at Imperial College London by the Applied Modelling and Computational Group since 2006. It has been used in improving the stability of harbour breakwaters many of which are constructed with concrete armour unit layer [13, 14] and catalyst pellets for fixed bed catalytic reactors [15–17]. In this chapter, a FDEM code, *Solidity*, is used to capture the influence of complex shape more accurately and simulate the particle packing using both deformable and rigid body versions of the code.

This chapter will cover code parallelisation performance and discussion of results showing both deformable and rigid body versions of the code in different application scenarios. *Solidity* also provides post-processing tools to analyse the particle packing structure in terms of local porosity and orientation distributions, contact forces, coordination number, etc.

The chapter is organised as follows: Section 8.2 briefly describes the mathematical models, Sections 8.3 and 8.4 show applications and some results, respectively, followed by conclusions.

8.2 Mathematical Models

The FDEM method was pioneered by Munjiza whose first working 2D FDEM code was developed in 1990 [18]. FDEM has proven its efficiency and reliability as a computational tool to solve problems involving transient dynamics of systems in which deformation and fracturing play an important role. Xiang et al. [19] introduced higher-order large strain improvements in their FDEM code, named Y3D-D. Y3D-D handles the transient dynamics of complex-shaped multi-body behaviour and can support a vast range of alternative e.g. non-linear constitutive or internally fracturing properties. Recently, Xiang developed a new version called Y3D-R to simulate rigid solid body interactions. The Y3D-R has the advantage that it executes the simulation faster than the general deformable version, Y3D-D.

8.2.1 Governing Equations

8.2.1.1 Y3D-D: Deformable Solid

For deformable solid, three classes of material models have been implemented in Solidity: elasticity, viscoelasticity, and plasticity. The motions of elements are influenced by the forces acting on elemental nodes, internal nodal force, external nodal force, $\mathbf{f}_{\text{ext}}$. The constitutive equation influences the deformation of the material through the Cauchy stress tensor in the linear momentum equation given by,

$$\mathbf{M}\dot{\boldsymbol{v}}_n^T + \mathbf{f}_{\text{int}} = \mathbf{f}_{\text{ext}} + \mathbf{f}_{\text{c}} \tag{8.1}$$

where $\boldsymbol{v}$ is nodal velocity vector. Internal nodal forces are given by

$$\boldsymbol{f}_{\text{int}} = \int_{v^{(n)}} \frac{\partial \boldsymbol{N}}{\partial \boldsymbol{x}} \boldsymbol{T} \, \mathrm{d}v \tag{8.2}$$

External nodal forces are given by

$$\boldsymbol{f}_{\text{ext}} = \int_{v^{(e)}} \boldsymbol{N}\boldsymbol{b} \, \mathrm{d}v + \int_{v^{(e)}} \boldsymbol{N}\boldsymbol{t} \, \mathrm{d}a \tag{8.3}$$

where $\boldsymbol{b}$ is body force, defined by $\boldsymbol{b} = \lfloor b_x \; b_y \; b_z \rfloor$ and $\boldsymbol{t}$ is surface traction force.

Nodal mass matrix is given by,

$$\boldsymbol{M} = \int_{V_0} \rho_0 \boldsymbol{N}\boldsymbol{N}^T \, \mathrm{d}v \tag{8.4}$$

Since mass is conserved, the mass is calculated based on the initial configuration. ρ is the density of material.

8.2.1.2 Y3D-R: Rigid Solid

When the solid bodies are assumed as rigid, they do not deform under external forces. To simulate the motion of rigid bodies in three-dimensional space, we have to extend Newton's second law to define linear and rotational motions.

$$m_i \ddot{\boldsymbol{x}}_i = \boldsymbol{f}_{d,i} + m_i \boldsymbol{g} + \sum_{j=1}^{k_i} (\boldsymbol{f}_{cn,ij} + \boldsymbol{f}_{ct,ij}) \tag{8.5}$$

$$I_i \ddot{\theta}_i = \sum_{j=1}^{k_i} \boldsymbol{T}_{c,ij} \tag{8.6}$$

where m_i and I_i are the mass and moment of inertia, respectively. $\ddot{\boldsymbol{x}}$ and $\ddot{\boldsymbol{\theta}}$ are the linear acceleration and angular acceleration of the particle i, respectively. $\boldsymbol{T}_{c,ij}$ is contact force torque. For multiple interactions, the interparticle forces and torques are summed for k_i particles interacting with particle i. $\boldsymbol{f}_{d,i}$ is the damping force given by Hart et al. [20],

$$\boldsymbol{f}_{d,i} = \alpha \left| \sum_{j=1}^{k_i} (\boldsymbol{f}_{cn,ij} + \boldsymbol{f}_{ct,ij}) \right| \operatorname{sgn}(\ddot{\boldsymbol{X}}_i) \tag{8.7}$$

where α is the coefficient of the mass damping.

8.2.2 Contact Force

In FDEM, a penalty function method is employed to calculate the normal contact force when two particles are in contact. The penalty function method in its classical form assumes that two particles penetrate each other. The elemental contact force is directly related to the overlapping area of finite element in contact. The distributed contact force approach considers the shape and the size of the overlapped volume for distribution among the surrounding nodes. The contact force is given by Munjiza [18],

$$\boldsymbol{f}_c = \sum_{i=1}^{n}\sum_{j=1}^{m}\int_{\Gamma_{\beta_{c_i}\cap\beta_{t_j}}} \boldsymbol{n}_{\Gamma_{\beta_{c_i}\cap\beta_{t_j}}}(\phi_{c_i} - \phi_{t_j})\,\mathrm{d}\Gamma \tag{8.8}$$

where β_c and β_t are the contactor and target discrete elements, respectively, $\boldsymbol{n}$ is the outward unit normal to the boundary of the overlapping area, the integration over finite element boundaries can be written as summation of integration over the edges of finite elements.

Xiang et al. [19] developed further the FDEM method by taking account of the sliding friction force. The well-known classic Coulomb type of friction was implemented and described as follows:

$$\boldsymbol{f}_t = -k_t\boldsymbol{\delta}_t - \eta\boldsymbol{v}_t \tag{8.9}$$

where η is the coefficient of viscous dissipation, $\boldsymbol{k}_t$ is the spring stiffness in tangential direction, $\boldsymbol{f}_t$ is the tangential elastic contact force and $\boldsymbol{v}_t$ is the tangential relative velocity.

If $\boldsymbol{f}_t$ is larger than the friction force obeying the Coulomb-type friction law, $|\boldsymbol{f}_t| > \mu\,|\boldsymbol{f}_n|$ the particles slide over each other and the tangential force is calculated using the total normal elastic contact force $\boldsymbol{f}_n$:

$$\boldsymbol{f}_t = -\mu\boldsymbol{f}_n \tag{8.10}$$

where μ is the coefficient of sliding friction.

8.2.3 Code Optimisation and Parallelization

High Central Processing Unit (CPU) cost hinders the FDEM codes from being widely used due to in-efficient and serial algorithms. Recently, Xiang optimised the contact detection algorithm in the FDEM and parallelized the code using Open Multi-Processing (OpenMP). In this chapter, the authors use a packing system with 288 rock-like boulders to test the CPU time. Figure 8.1 shows the FDEM capability to model such a system of mono-sized angular rock-like boulders (bodies of about 40 kg) being dumped in a 5.26 m × 5.3 m × 3.87 m bin (front face not shown). The OpenMP parallelized code is shown to give a speedup of 6.5 on 8 threads and 9 on 12 threads in a 3D deformable rock deposition with 288 particles in Figure 8.1. The runtime is reduced by half after the code was optimised. In future, it is planned for hybrid message passing interface (MPI) and OpenMP to be implemented for large particle systems (Figure 8.2).

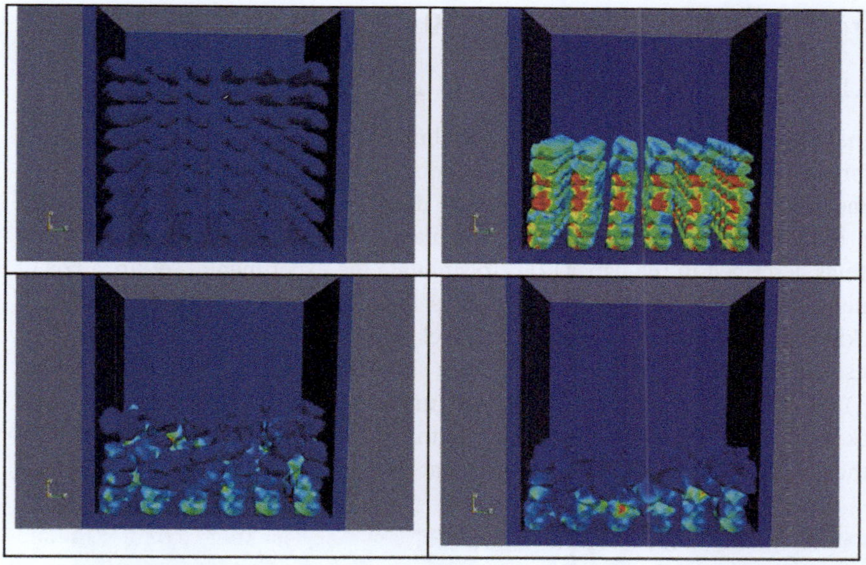

Figure 8.1 Three hundred rocks depositing in a container, element number: 18845 Node number: 9540 Contact number 0-9000.

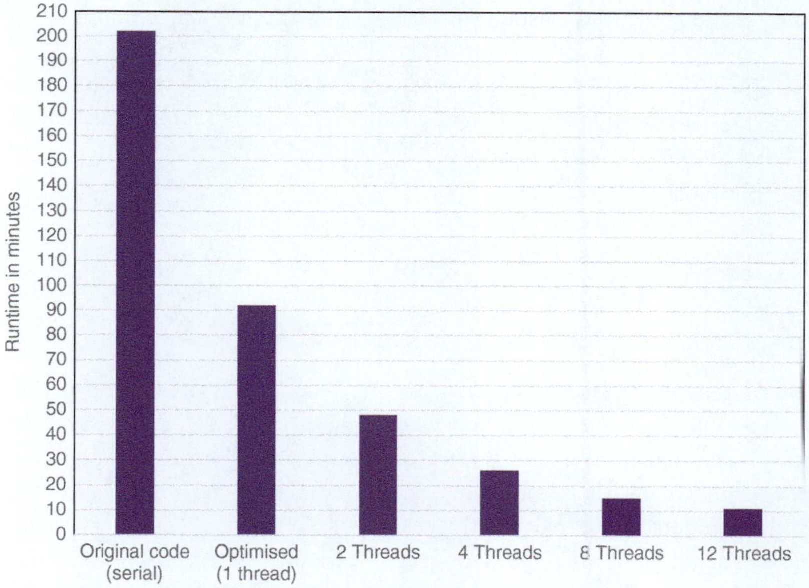

Figure 8.2 Speedup achieved by the Solidity, Workstation specification: Intel(R) Xeon(R) CPU X5680 @ 3.33 GHz cache size: 12 Mb, maximum threads number: 12.

8.3 Application of Cube Packing

8.3.1 Numerical Simulation Set-up

Latham and Munjiza [21] designed and built a cube-packing rig. They used a deformable FDEM code to simulate the movements of the cubes and compared numerical results with the experimental data. In this study, both deformable and rigid FDEM codes were used to simulate the cube packing using the same settings described in the paper [21]. To simulate the experimental set-up, a box of dimensions 250 mm × 250 mm × 375 mm is used. A random sequential addition packing protocol is used to prepare the packing of 648 identical cubes in the box. These cubes are deposited in spaced array sheets sequentially at a certain height (h_{de}) from the domain base under gravity until the distance between the h_{de} and pack surface is less than the maximum length between particle surface and centroid of the particle (Figure 8.3). This ensured that the particles land with an appropriate

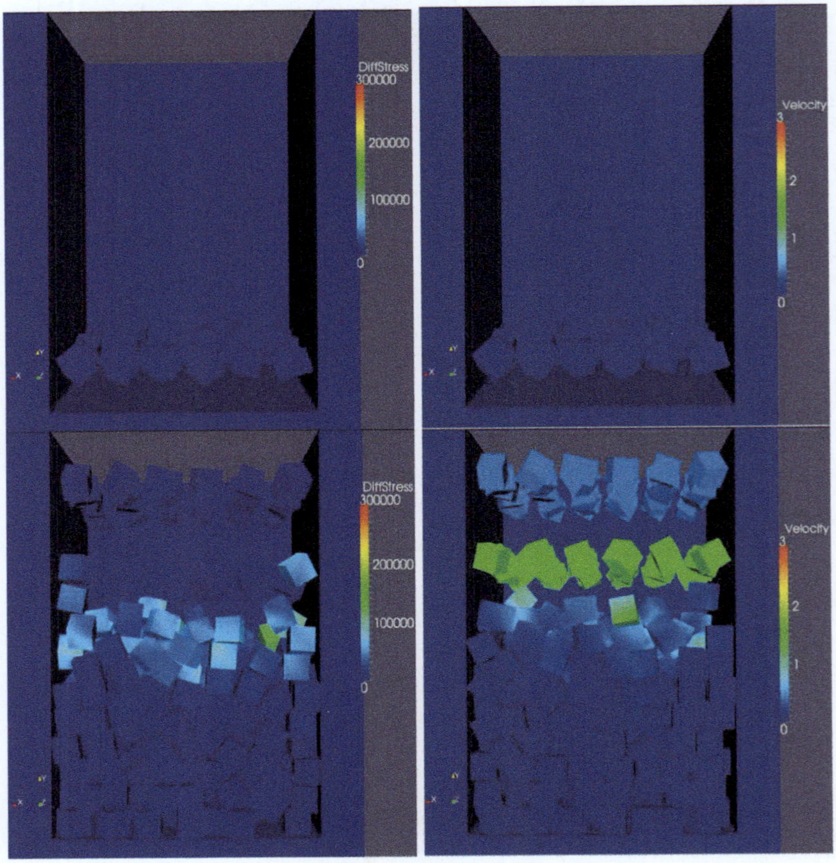

Figure 8.3 Left column: deposition sequence for cube packing using deformable version of FDEM code, the colours represent the magnitude of differential stress applied on particle; right column: deposition sequence for cube packing using rigid version of FDEM code; the colours represent the magnitude of the particle velocity, the hotter the faster they are moving.

momentum comparable to that in experiments and without significantly moving the particles which were already in place in the packing. The cubes are modelled with a range of Coulomb coefficients of friction from 0.0 to 1.0 and as bodies with density of 0.653 $g\,cm^{-3}$, Young's modulus, 1 GPa, Poisson's ratio, 0.3, and a damping coefficient of 0.6.

8.3.2 Results and Discussion

To compare the numerical results between deformable and rigid FDEM codes, the same random deposition protocols and the same material density for box and cubes were used. However, some differences in damping methods are used. This is because in the deformable i.e. viscoelasticity model, viscous damping is used, but in the rigid FDEM code, mass damping is used. The two parameters required are not the same, although both are related to kinetic energy dissipation. The final packing structure is very sensitive to the mass damping term; hence, several trial tests have been carried out to calibrate the mass damping coefficient. Figure 8.4 clearly shows the simulation using the rigid FDEM code can be calibrated and then shown to match the one using the deformable code although the simulation does not match exactly, particle for particle.

After calibrating the mass damping coefficient, we used the FDEM rigid version to investigate the effect of the friction coefficient on packing structure. The friction coefficient was changed from 0.0 to 1.0, and six numerical tests were carried out. The snapshots of the deposition of the cube packing are shown in Figure 8.5. It shows that a strong wall effect occurs when the friction coefficient is small. When the friction coefficient is zero, the cubes tend to be rearranged in a regular pattern and aligned along the walls. When the friction coefficient increases, the cube packing becomes more 'chaotic'. It was also observed that the smaller the friction coefficient, the longer time for the particles to settle in the container. This means that

(a) (b)

Figure 8.4 Simulation of deposition leading to a packed structure of cubes with friction coefficient 0.38, using (a) deformable FDEM code; (b) rigid FDEM code.

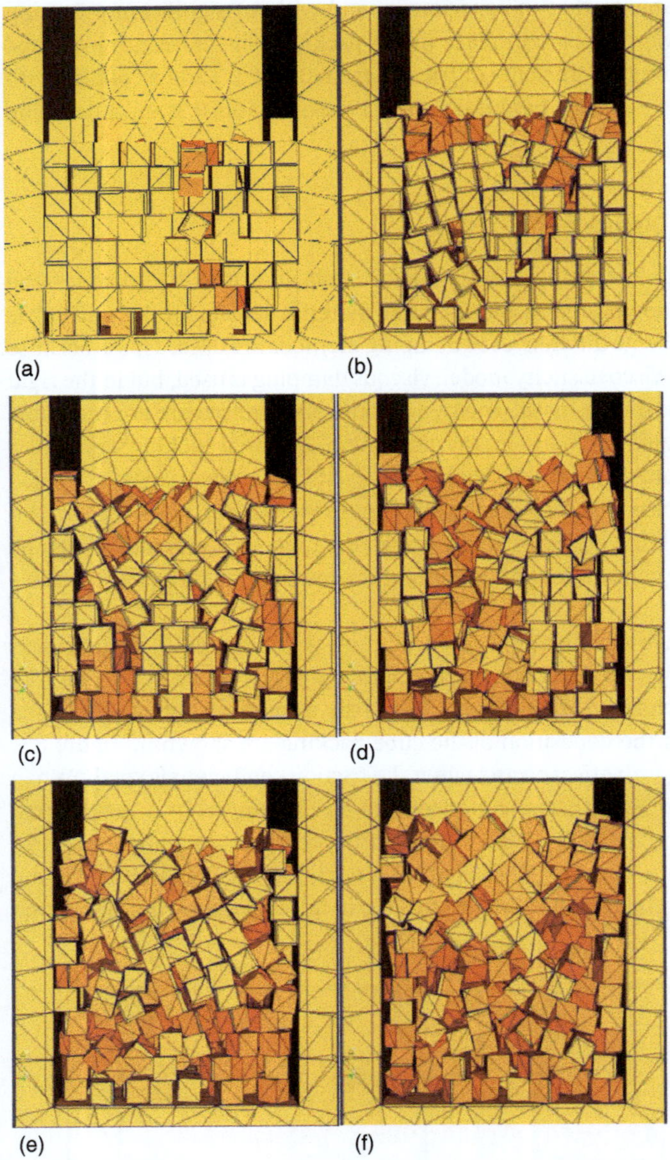

Figure 8.5 Numerical simulation of cube packing for friction coefficient (a) 0.0, (b) 0.1, (c) 0.2, (d) 0.3, (e) 0.5, (f) 1.0. Figures represent the front face view with the box side removed.

when the friction coefficient is small, the kinetic energy is dissipated more slowly, which can be used to rearrange the cubes into a tighter and regular pattern.

Figure 8.5 shows that there is a wall effect in several packings. To qualitatively study the wall effect three sample windows of three different dimensions (see Figure 8.6) are selected to calculate average packing density: 250 mm × 250 mm ×

Figure 8.6 Sample window for packing density calculations.

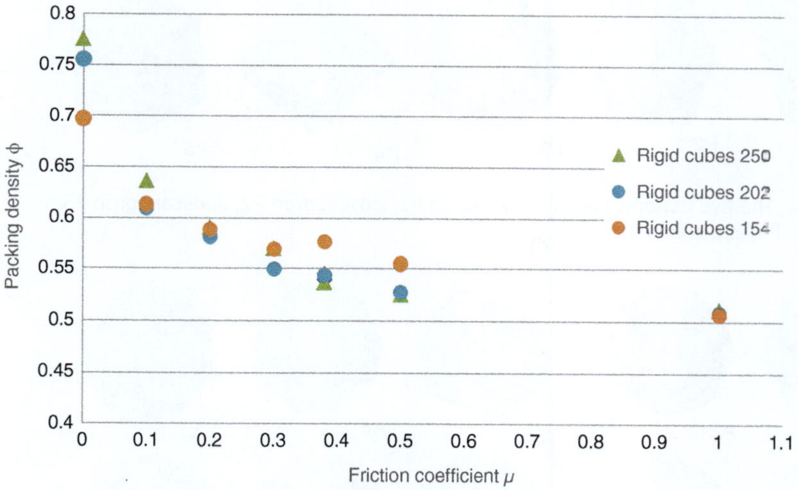

Figure 8.7 Packing density of cubes at rest.

250 mm, 202 mm × 202 mm × 202 mm, 154 mm × 154 mm × 154 mm. We calculated the average packing densities for all cube packings with different friction coefficients and FDEM-rigid version. This shows that friction coefficient has a great influence on the packing density. Figure 8.7 shows the effects of friction coefficient μ on packing density. As expected, the higher the friction coefficient, the lower is the resulting packing density. It can be seen from Figure 8.7 that very strong wall effects exist in several packings. For $\mu = 0.0$, censoring the wall zone leads to lower packing densities because the orthogonal tower stacking is especially well developed at the wall, whereas for $\mu = 0.38$, and 0.5, the central part has the higher density suggesting the wall region inhibits tighter packing at these more realistic friction values for wooden cubes.

8.4 Application of the Packing of Platonic and Archimedean Solids

8.4.1 Numerical Simulation Set-up

The same random sequential addition packing protocol which was described in Section 8.3.1 is used to prepare the packing of 5 Platonic (Figure 8.8) and 13 Archimedean (Figure 8.9) solids in a rectangular container. About 11 000 particles are added sequentially (in spaced array particle sheets) at a certain height from the container base until the distance between the depositing height and pack surface is less than the maximum length between particle surface and the particle centroid. This ensured that the particles land in a stable configuration without significantly moving the particles which were already in place in the packing. In this chapter, Coulomb coefficients of friction from 0.2 to 1.0 are considered and the

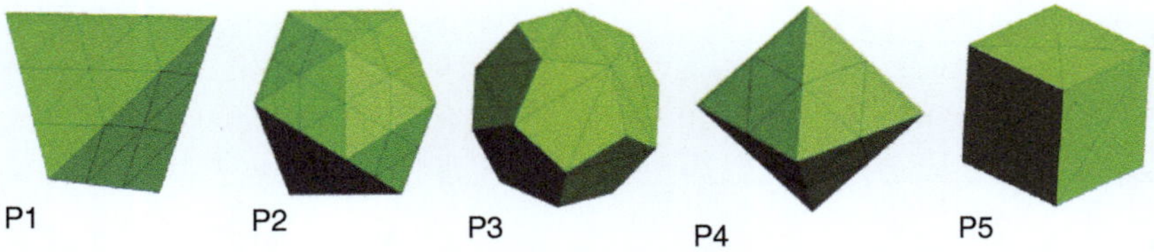

Figure 8.8 The five Platonic solids: tetrahedron P1, icosahedron P2, dodecahedron P3, octahedron P4, and cube P5.

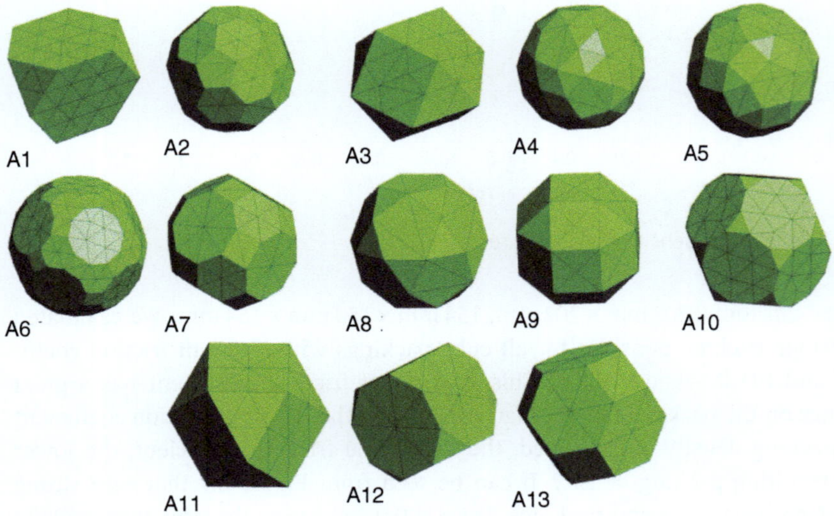

Figure 8.9 The 13 Archimedean solids: truncated tetrahedron A1, truncated icosahedron A2, snub cube A3, snub dodecahedron A4, rhombicosi dodecahedron A5, truncated icosidodecahedron A6, truncated cuboctahedron A7, icosidodecahedron A8, rhombicub octahedron A9, truncated dodecahedron A10, cuboctahedron A11, truncated cube A12, and truncated octahedron A13.

particles are modelled as rigid bodies with density of 2.56 g cm^{-3}, and the damping coefficient, 0.6.

8.4.2 Results and Discussion

8.4.2.1 Packing Density

Porosity in a granular system is the volume fraction of voids in a total volume of a granular system, the volume fraction of solids being the packing density. As seen in Figure 8.7, the contained walls have a great influence on the packing density. The packing density tends to decrease near the packing surface, this is sometimes called the surface effect in packing studies. To reduce these two effects, we select a smaller sample box (5d away from container walls and packing surface) to calculate the average packing density (Figure 8.10).

Five Platonic (P1-5) and six Archimedean (A1, A3, A5, A7, A10, A13) solids are investigated at this stage. They are deposited in the container using a sequential addition protocol with friction coefficient 0.6. The mean packing densities are shown in Table 8.1. These results demonstrate that whereas friction has quite a strong effect on packing density, the packing densities for all 11 solids vary in a narrow range, 0.52–0.57 when a constant realistic friction is chosen for all shapes. Truncated tetrahedron A1 is selected for the effect of friction.

The effects of friction coefficient μ on packing density is shown in Figure 8.11. As expected, the higher the friction coefficient, the lower is the packing density

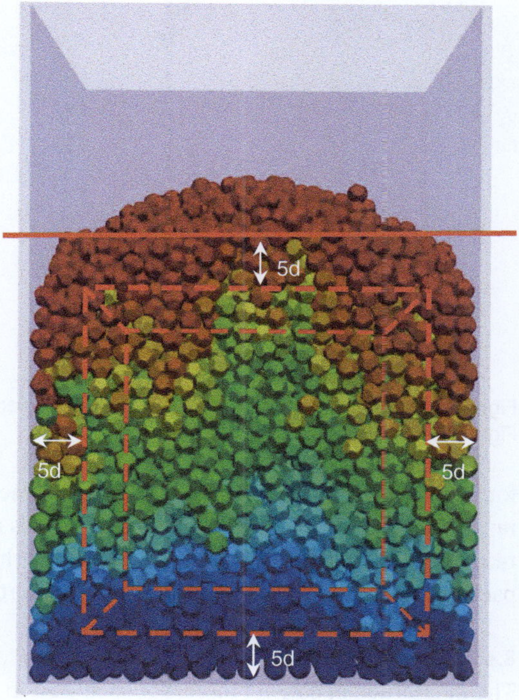

Figure 8.10 Final simulated packing structure for the truncated dodecahedron.

Table 8.1 Packing density.

	Packing density
Truncated tetrahedron A1	0.57
Snub cube A3	0.55
Rhombicosi dodecahedron A5	0.57
Truncated cuboctahedron A7	0.56
Truncated dodecahedron A10	0.54
Truncated octahedron A13	0.56
Tetrahedron P1	0.52
Icosahedron P2	0.56
Dodecahedron P3	0.56
Octahedron P4	0.55
Cube P5	0.52

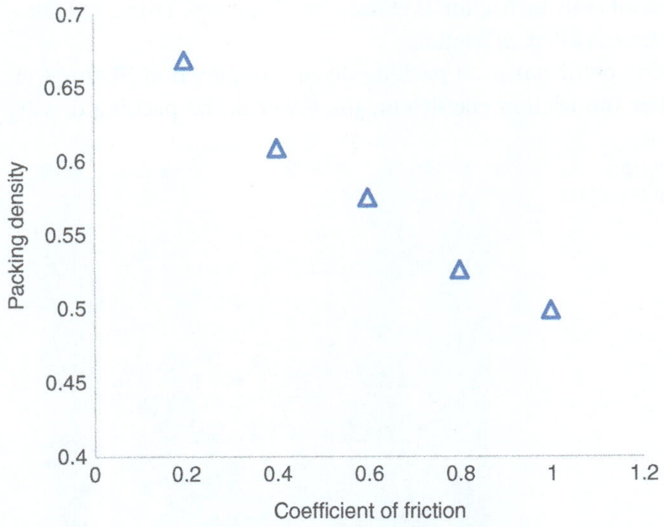

Figure 8.11 Packing density versus friction coefficient for truncated tetrahedron A1, see Figure 8.9.

for the A1 shape. This trend qualitatively agrees with the results found by other researchers [22]. When the friction coefficient is reduced to 0.2, the packing density reaches 0.67. This value matches very well with the density of the optimal lattice packing of the truncated tetrahedron, 0.68 reported in [2].

8.4.2.2 Contact Forces

The magnitude of contact forces for each unit in the static pack has been extracted. The maximum contact force experienced by each particle is also of interest.

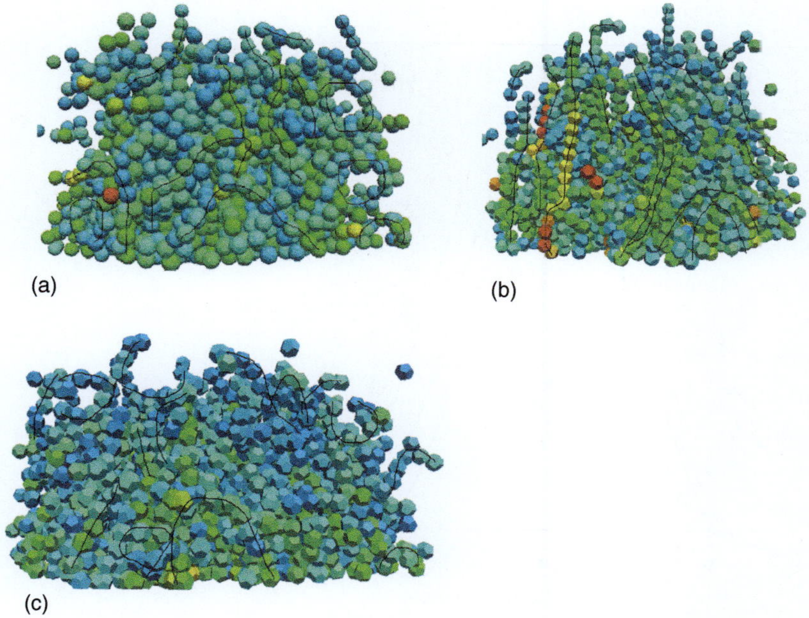

Figure 8.12 Packing structures for (a) rhombicosi dodecahedron A5, (b) truncated dodecahedron A10, (c) dodecahedron P3. Colours represent the magnitude of the maximum contact force experienced by each particle, the force chain networks (black solid lines) link all particles with maximum contact forces greater than 1.5 times the mean maximum contact force.

Normally, the contact force is used to extract the force-chain network. In this chapter, the force chain network is constructed from a list of all contact forces which are greater than 1.5 times the mean maximum contact force. A class of dodecahedron, P3, A5, and A10 is selected for analysing the structure pattern (Figure 8.12a–c). It is shown that A10 tends to generate long and straight chains due to bigger facets. P3 and A5 tend to form short and circular networks. Arch-type structures are found in all three packs near the corners of the container.

8.4.2.3 Coordination Number

The number of contacts each particle makes with its neighbours is known as the coordination number. The coordination number has a wide variation; contacts with neighbouring particles or the container ranging from 1 to 10 in the three packs examined for A5 A10 and P3. Different shapes create packing conditions resulting in a different average coordination number and different spatial distributions of coordination number. Figure 8.13 shows the variation of coordination number in a probability density plot for the three shapes with different packs. From A10 to P3, the probability density curve shifts to the right, resulting in a larger average coordination number. It is worth noting that the analysis tools recognise the surface-to-surface and edge-to-surface contacts as one coordination number. Thus, the lowest value of coordination number in these three static packs is one. The analysis tools including

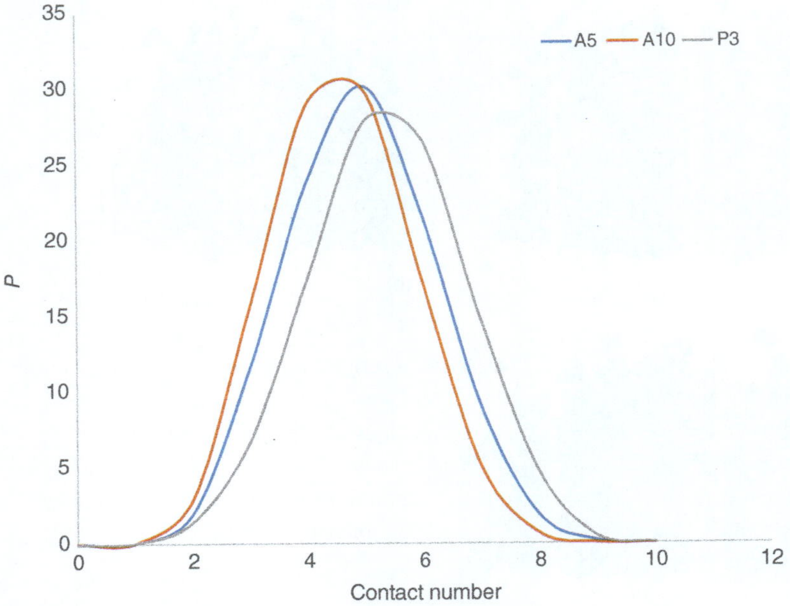

Figure 8.13 Probability density function, P of normalised contact number.

force- and stress-chain statistics are continually being improved for the benefit of future granular media analysis.

8.5 Conclusions

In this chapter, a FDEM model was developed and parallelised to create efficient faster simulations to investigate the random packing of complex shapes.

We used both Y3D-D and Y3D-R to simulate cube packing with different friction coefficients. The effect of the friction for cubes was investigated. In natural systems, potential or kinetic energy available for rearrangement of particles and tighter packing is absorbed by friction, stress wave propagation and in-elastic material damage. Both Y3D-D and Y3D-R use absorption coefficients to help account for the in-elastic damage via a viscous damping, but only Y3D-D captures the kinetic energy associated with stress wave propagation.

Platonic and Archimedean body packs were investigated and analysed in terms of local porosity, maximum contact forces, coordination number, etc. The effect of friction was investigated for truncated dodecahedron particles. Packing density decreases as friction coefficient increases. The sequential addition method of introducing the particles to the already packed grains was set up to give only low levels of energy and with the one friction value tested in this work, relatively low values of packing density will have been created compared with maximum jammed packs that would be possible with different coalescence history and friction parameter selection. The packing density analysis indicates that slightly lower packing densities will be produced for the most angular and least spherical particle shapes which were the tetrahedrons and cubes.

Acknowledgements

The authors' gratefully acknowledge the support of EPSRC and Johnson Matthey.

References

1 Cundall, P.A. and Strack, O.D.L. (1979). A discrete numerical model for granular assemblies. *Géotechnique* 29 (1): 47–65.

2 Torquato, S. and Jiao, Y. (2009). Dense packings of the Platonic and Archimedean solids. *Nature* 460 (7257): 876–879.

3 Caulkin, R., Jia, X., Xu, C. et al. (2009). Simulations of structures in packed columns and validation by X-ray tomography. *Industrial and Engineering Chemistry Research* 48 (1): 202–213.

4 Favier, J.F., Abbaspour-Fard, M.H., and Kremmer, M. (2001). Modeling nonspherical particles using multisphere discrete elements. *Journal of Engineering Mechanics* 127 (10): 971–977.

5 Garcia, X., Latham, J.-P., Xiang, J., and Harrison, J.P. (2009). A clustered overlapping sphere algorithm to represent real particles in discrete element modelling. *Géotechnique* 59 (9): 779–784.

6 Zhong, W., Yu, A., Liu, X. et al. (2016). DEM/CFD-DEM modelling of non-spherical particulate systems: theoretical developments and applications. *Powder Technology* 302: 108–152.

7 Cleary, P., Stokes, N., and Hurley, J. (1997). *Efficient Collision Detection for Three Dimensional Super-ellipsoidal Particles*, 139–144. Singapore: World Scientific.

8 Delaney, G.W., Morrison, R.D., Sinnott, M.D. et al. (2015). DEM modelling of non-spherical particle breakage and flow in an industrial scale cone crusher. *Minerals Engineering* 74: 112–122.

9 Greben, J.M., Gledhill, I., Cooper, A.K., de Villiers, R. (2010). Characterization and Properties of Breakwater Structures modelled by a Physics Engine. 7th South African Conference on Computational and Applied Mechanics (SACAM10), University of Pretoria, Pretoria, 10–13.

10 Moghaddam, E.M., Foumeny, E.A., Stankiewicz, A.I., and Padding, J.T. (2019). Fixed bed reactors of non-spherical pellets: importance of heterogeneities and inadequacy of azimuthal averaging. *Chemical Engineering Science: X* 1: 100006.

11 Moghaddam, E.M., Foumeny, E.A., Stankiewicz, A.I., and Padding, J.T. (2018). Rigid body dynamics algorithm for modeling random packing structures of nonspherical and nonconvex pellets. *Industrial and Engineering Chemistry Research* 57 (44): 14988–15007.

12 Huang, H. and Tutumluer, E. (2014). Image-aided element shape generation method in discrete-element modeling for railroad ballast. *Journal of Materials in Civil Engineering* 26 (3): 527–535.

13 Latham, J.-P., Anastasaki, E., and Xiang, J. (2012). A FEMDEM numerical model study of rubble-mound structures armoured with concrete armour units. *Coastal Engineering Proceedings* 1 (33):posters.37.

14 Latham, J.-P., Xiang, J., and Higuera, P. (2015). Numerical modelling of the stability of breakwater armour systems. In: *Coastal Structures and Solutions to Coastal Disasters*, 687–698. American Society of Civil Engineers https://ascelibrary.org/doi/abs/10.1061/9780784480304.073.

15 Farsi, A., Xiang, J., Latham, J.-P. et al. (2021). Packing simulations of complex-shaped rigid particles using FDEM: an application to catalyst pellets. *Powder Technology* 380: 443–461.

16 Farsi, A., Xiang, J., Latham, J.-P. et al. (2017). Simulation and characterisation of packed columns for cylindrical catalyst supports and other complex-shaped bodies. In: *Proceedings of the 7th International Conference on Discrete Element Methods*, Springer Proceedings in Physics, vol. Vol. 188 (ed. X. Li, Y. Feng and G. Mustoe), 397–406. Singapore: Springer Singapore http://link.springer.com/10.1007/978-981-10-1926-5_42.

17 Latham, J.-P., Xiang, J., Farsi, A. et al. (2020). A class of particulate problems suited to FDEM requiring accurate simulation of shape effects in packed granular structures. *Computational Particle Mechanics* 7 (5): 975–986.

18 Munjiza, A. (2004). *The Combined Finite-Discrete Element Method*. Wiley https://onlinelibrary.wiley.com/doi/abs/10.1002/0470020180.ch1.

19 Xiang, J., Munjiza, A., and Latham, J.-P. (2009). Finite strain, finite rotation quadratic tetrahedral element for the combined finite-discrete element method. *International Journal for Numerical Methods in Engineering* 79 (8): 946–978.

20 Hart, R., Cundall, P.A., and Lemos, J. (1988). Formulation of a three-dimensional distinct element model—part II. Mechanical calculations for motion and interaction of a system composed of many polyhedral blocks. *International Journal of Rock Mechanics and Mining Science and Geomechanics Abstracts* 25 (3): 117–125.

21 Latham, J.–.P. and Munjiza, A. (2004). The modelling of particle systems with real shapes. In: *Philosophical Transactions of The Royal Society a Mathematical Physical and Engineering Sciences*, vol. 362 (1822) (ed. K. Richards and M. Dove), 1953–1972.

22 Zhao, S., Zhou, X., Liu, W., and Lai, C. (2015). Random packing of tetrahedral particles using the polyhedral discrete element method. *Particuology* 23: 109–117.

Index

Simulations in Bulk Solids Handling: Applications of DEM and other Methods, First Edition.
Edited by Don McGlinchey.

d